Praise for
The Garden Tool Handbook

Holy Dibbler! Tools I never knew existed so I never knew I needed them. A fascinating book, fabulously illustrated and a great tool itself!

—Jeff Lowenfels, North America's longest running columnist and author of the Teaming book series

What a surprise! I was expecting a great review of tools including the many options that are available, and *The Garden Tool Handbook* does deliver on that, but more importantly it connects tools to the gardening process. It breaks the market gardening process into stages from indoor seed starting to final harvesting and then helps the reader select the right tool and accessory for each step in the process. You will also discover fun tidbits of information smattered throughout the book ranging from historical anecdotes to special care techniques. It even shows you how to modify commercial tools to make them more productive, and how to make some yourself to reduce costs. If you are serious about growing vegetables, this book will save you a huge amount of time.

—Robert Pavlis, author of the *Science for Gardeners* book series and *Garden Myths*

Another superbly practical book from Zach Loeks! He covers the season in 16 stages, helping you make wise choices when designing and gathering a tool system for each stage, suiting your operation and scale.

Well-lit photos show the variations on each type of tool, and you will likely meet some new to you. I'm very enamored of the manual tine-weeder and the homemade leek dibbler! More complex tools have photos with numbered parts and clear instructions.

Tools extend our human power, reach, and capacity. Good tools, well understood, make for a better life. The right size rubber band for the crop will save frustration.

—Pam Dawling, author of *Sustainable Market Farming* and *The Year-Round Hoophouse,* speaker, teacher, consultant, and vegetable grower

One of the challenges getting into gardening or farming is simply knowing what tools are actually out there and how to distinguish between so many that might look the same. Zach Loeks lays out the full range of tool options for all the different gardening operations you might undertake, and walks you through when you might want to use each of them.
—Dan Brisebois, author, *The Seed Farmer* and host, the Seed Farmer podcast

Praise for Other Titles by Zach Loeks

The Two-wheel Tractor Handbook

This book is truly unique, full of tips, hints, and tidbits even for someone like me who's been farming with such tractors for over 20 years.
—Jean-Martin Fortier, author, *The Market Gardener*

Jam-packed with great information. This book is going to save serious growers a lot of time tinkering around.
—Curtis Stone, author, *The Urban Farmer*

This book discusses scale and choosing a family of machinery that work together and allow a farm to grow over time in a comprehensive and innovative way that I have not seen anywhere else.
—Sam Oschwald Tilton, vegetable farming mechanization specialist

The Edible Ecosystem Solution

Zach Loeks uses a rich palette of strategies and species that make sense in his region, to inspire people everywhere to create an edible ecosystem in theirs, and so become more connected to and nourished by nature.
—David Holmgren, co-originator of permaculture

The Edible Ecosystem Solution is at once a work of art and a valuable tool in our journey towards an ecological society.
—Darrell E. Frey, Three Sisters Farm, author, *Bioshelter Market Garden*, co-author, *The Food Forest Handbook*

The Permaculture Market Garden

Bringing permaculture's holistic thinking to the problems of market farming, Zach Loeks has done this burgeoning economic sector a world of good with his pithy words and lovely drawings. Business planning was never before this colorful, soulful, or needed for the times.

—Peter Bane, author, *The Permaculture Handbook*

A legacy of grateful land is surely a memorial that can capture both our imagination and our physical effort. Engaging in this discovery and planning process is worth the effort, and Zach gives us another tool to engage more strategically. Now get out paper, pencil, ruler, and graph paper. It's that simple, and that rewarding.

—Joel Salatin, Polyface Farm

THE GARDEN TOOL HANDBOOK

FOR SERIOUS GARDENERS TO SMALL SCALE FARMERS

ZACH LOEKS

**Copyright © 2025 by Jedediah Loeks.
All rights reserved.**

Cover design by Diane McIntosh.

Cover image: Zach Loeks.

Printed in Canada. First printing November 2024.

This book is intended to be educational and informative. It is not intended to serve as a guide. The author and publisher disclaim all responsibility for any liability, loss, or risk that may be associated with the application of any of the contents of this book.

Inquiries regarding requests to reprint all or part of *The Garden Tool Handbook* should be addressed to New Society Publishers at the address below. To order directly from the publishers, please call 250-247-9737 or order online at www.newsociety.com.

Any other inquiries can be directed by mail to:
New Society Publishers
P.O. Box 189, Gabriola Island, BC V0R 1X0, Canada
(250) 247-9737

LIBRARY AND ARCHIVES CANADA CATALOGUING IN PUBLICATION

Title: The garden tool handbook : for serious gardeners to small scale farmers / Zach Loeks. Names: Loeks, Zach, 1985- author

Description: Includes index.

Identifiers: Canadiana (print) 20240468775 | Canadiana (ebook) 20240468783 | ISBN 9781774060025 (softcover) | ISBN 9781550927955 (PDF) | ISBN 9781771423915 (EPUB)

Subjects: LCSH: Garden tools—Handbooks, manuals, etc. | LCSH: Gardening—Handbooks, manuals, etc. | LCSH: Farms, Small—Handbooks, manuals, etc. | LCGFT: Handbooks and manuals.

Classification: LCC SB454.8 .L64 2025 | DDC 635—dc23

New Society Publishers' mission is to publish books that contribute in fundamental ways to building an ecologically sustainable and just society, and to do so with the least possible impact on the environment, in a manner that models this vision.

Contents

PART 1: THE TOOLS 1
- The Beginning 1
- Extending Human Power 4
- All about Tools 11
- Tool Types 18
- Tool Design 22
- DIY or Buy 35
- Garden Operation Cycle 44
- Tool System Design 51
- Scale-Based Decision-Making 56
- Scaling-Up Tool Systems 61
- Scaling-Up Your Avatar 64

PART 2: YOUR GARDEN OPERATION CYCLE 68
- **Chapter 1:** Site Analysis and Sampling 68
- **Chapter 2:** Garden Design and Crop Planning 74
- **Chapter 3:** Garden Starts 81
- **Chapter 4:** Primary Land Preparation 90
- **Chapter 5:** Plot and Permabed Forming 97
  - Case Study: Container Gardening 102
- **Chapter 6:** Fine Seedbed Preparation 108
- **Chapter 7:** Seeding and Planting 116
- **Chapter 8:** Irrigation 134
- **Chapter 9:** Garden Crop Maintenance 144
- **Chapter 10:** Crop Weeding 164
- **Chapter 11:** Garden Harvest 176
- **Chapter 12:** Season Extension 180

Chapter 13: Post-harvest Handling and Curing	184
Chapter 14: Cold Storage	191
Chapter 15: Marketing	194
Chapter 16: Cleanup and Maintenance	198
Conclusion	221
Resources	222
List of Tool Companies and Suppliers	222
Index	224
About the Author	231
Praise	231
About New Society Publishers	232

Part 1: The Tools

The Beginning

I like tools.

Tools have always been interesting to me. Growing up on a Permaculture homestead, we always had spades, shovels, digging forks, post-hole diggers, rakes, and trowels in the shed. I loved to explore our wellhouse and see the pump, pressure tank, and manifolds that sent water around our property. At my grandmother Deva's homestead in Sebastopol, California, I loved to see her tool buckets and her beloved garden journal (I still flip through its rose-scented pages). In my own start-up market garden, I gleefully strode to those *early-days'* fields with my essential tools: stirrup hoe, garden rake, spade, shovel, and wheel hoe, ready to do weed sweeps in the bean rows and prepare new beds for summer salad greens. I cruised up and down rural fence lines planting trees with a sturdy spade for many years—hand planting over 100,000 saplings on my and other rural properties as part of native woodland restoration, riparian buffer strip, and upland shelter belt projects for farms. As my market farm scaled-up, I gained new tools and equipment, but I continued to use practical favorites. I have forked more garlic and carrot beds than anyone would care to remember while gaining a thorough education in classic literature, good fiction, and awesome tunes on my ear buds while harvesting for a 300-member CSA, two urban farmer's markets, and on-farm sales. My seed garlic business burgeoned; as a pro grower, I grew over 100,000 heirloom garlic bulbs and over 125 different varieties each year—one of the most diversified online seed garlic sellers in Canada. It was for this endeavor that I designed a five-row DIY dibbler to improve accuracy, uniformity, and efficiency for fall planting of the cloves.

I have shifted my work to innovative design, research, and installation of food forests and other edible ecosystems as organized diversified polycultures on my farm and for community projects, and I continue to do trials and experiments with tools to see which ones best serve the needs of different types of growers (home, market, Permaculture, etc.) at different scales and in different contexts (urban, suburban, rural). I like tools, and I hope this book encourages the same fondness in you. It is my intent to help you make strategic purchases and inspire you to build your own DIY tools whenever they are needed to get ahead of the weeds, master the harvest, and much, much more!

When we hold garden tools, there is a feeling of being armed for "right," prepared to grow local, healthy, and sustainable food for ourselves, our family, and our community.

What Is This Book About?

This is a no-nonsense, step-by-step handbook to systematic growing—from crop planning to harvest, and beyond. This book takes a **tool's-eye view**, telling the story of the garden season from the point of view of the tools you need to be successful. After the introductory sections, the book outlines a standard **seasonal operation cycle** broken into **16 major production stages**, presenting tools and techniques for critical garden tasks within each stage. The techniques given are tried-and-true best management practices that work with different tool systems at different scales for different enterprises.

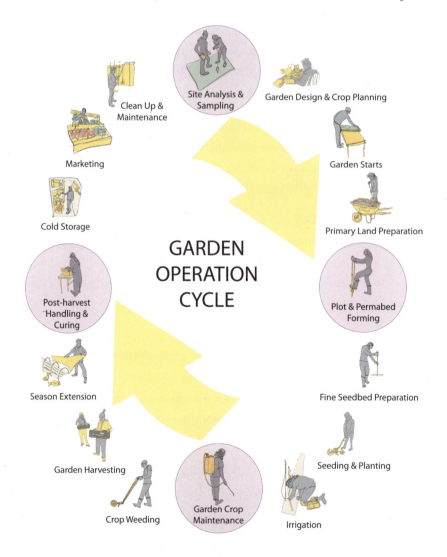

For All Growers

This book is for the home gardener, market grower, homesteader, or small-scale farmer. Whether you grow 1,000 ft^2 or 3 acres, it will help you choose tools wisely as you ***start-up***, ***scale-up***, and, finally, ***pro-up*** your production. These three **scale phases** are ways of thinking about where you stand in your enterprise's growth process—and which tools might mend the weak links in your production stages. Solving weak links with sound decision-making based on scale, soil, and best practices will help you assemble your own **complete tool system** for each production stage and meet all your garden tasks with ergonomic efficiency and cost-effective resilience. Having the right tool at hand makes a world of difference (and the wrong one, a world of trouble). You'll decide for your context, and this book is a guide for those decisions by showcasing a wide array of tools and considerations.

Extending Human Power

Humans are the original tool! We have always used our hands to engage with the natural world, plucking seeds, pressing soil, pulling at debris, and gleaning its ripening fruits. Our hands and bodies were the first contact with food plants millennia ago. Then we learned to fashion wood, bone, sinew, and stone to maximize our power, reach, and ability. *Tools are extensions of our hands, bodies, and minds*—performing garden tasks with **greater power** and **efficiency**. We then made tools out of bronze, iron, and steel. In the modern era, we further extended human power with power tools and more ergonomic designs. Through the millennia, tools used in farming have been designed and redesigned countless times to do essentially the same things: turn, drag, and mix soil; burn, bury, or pull material; and open, close, or pack soil, etc. Is this not the same work of our garden operations today—and our farms at any scale?

Long Ago on a Journey

Long ago, maybe someone on a hillside field picked up a sun-cured limb from the ground. Maybe they used it to herd feral sheep along a ridgeline. Stopping in the shade of an apple tree, they enjoyed its fruit and put some in their leather bag for later. Further on, they ate more apples and curiously turned a seed-dotted pome core in their hand. Then, with the pointed earthworn tip of their staff, they shoved it into the soil here, and there, and way over there as well. Into these dibbled holes, they placed the apple seeds (or perhaps they were from cucumber or squash…). Then turning their staff over, they pulled the crook end along the surface of the ground to cover the seeds in their depressions and patted the soil firm, watering it in from their bag and walking on to the next spring-fed seep to refill their bag and water the flock. When returning along this route weeks later, the person sees young plants growing and identifies the leaves of their favorites, and, pulling the staff across the soil around the emerging food plants, hacks the other vegetation down. This wooden stave may well have been *the first garden tool*, and a multifunctional one at that: shepherd crook, dibbler, trowel, spade, and hoe.

Or maybe, in a grass field somewhere, a cured animal jawbone was lashed to a wooden staff and used to hoe a squash field. And, along a stream at the base of a hill, long cuts in the ground were made to allow water to flow via gravity to the fields of squash. In a not-so-distant future, that once ancient squash garden could still be grown and picked by growers today, placed into crates, and loaded onto a cart for a farmer's market. Tools have always been used as ways of extending our reach, enhancing our power, and fine-tuning our techniques for growing food. Today, our tools are fine-tuned to perform an array of specialized garden tasks.

Top left: *Tools similar to these from Guatemala are found around the world as essential land preparation equipment. From left to right: a pick to open new stony ground, a broad hoe to form beds and trenches, a rake to finish bed tops, and a hand hoe for close action.*

Bottom left: *Wishing to tune into the uses of tools at the transition between hunting/gathering and more sedentary agriculture, I walked across a portion of the Anatolian highlands in Turkiye. Shown here is the oak stave I cut from a managed coppice in the hills. I traveled along shepherd trails, collecting seeds and sowing them into soil opened with the stave dibbler. I walked solo, but, historically, whole communities would have moved across the land at this time of year, settling only temporarily and leaving the land with more food varieties growing to return to when they passed through again.*

TOOL VIEW: Humans

Humans are born with tools: our hands and fingers, our feet and legs, our minds and spines. We are well-equipped in our human body to work soil, plant, harvest, and process food. You can reduce wear and tear on your original tools by employing fabricated tools (and remembering to stretch daily!).

FARM FEATURE
Life as Tools at the Edible Biodiversity Conservation Area

Are humans the original tool? Really, *all* of life can be seen as "tools" in the garden of Earth. This is an important mindset. Living organisms and our environment are important allies in productivity. We can bring compost to the garden with a wheelbarrow, but it is animal manure, rock minerals, and rotted plant debris that made that compost! For food growing, our primary tools are the soils (including the organisms that live in it) and the plants and animals that live on the farm. Climate, geology, and ecology equip us for success. When we remember these are the fundamentals of production, we can design farms to make use of these natural resources.

At the **Edible Biodiversity Conservation Area** (EBCA), plant diversity is our tool kit. Here, we are researching (A) many varieties of **edible and useful** perennials and annual crops for their potential to maximize ecosystem services. Our successful varieties are grown in **nursery plots** (B), where soil is a key tool. We improve it with cover crops and composted manure, we inoculate it with soil life, and we strive to conserve soil structure with low tillage practices. We employ **ecological design principles** (more fully outlined in my book, *The Edible Ecosystem Solution*).

These plots yield saplings for use in education sites within the EBCA and in nearby communities.

We grow mulberries, Asian pears, pecans, **hazelnuts** (C), and many other crops. We are also growing intensive coppices of hardwoods for our own tool handles, and **willows** (D) for biofuels, **basketry** (E), and living fence designs. Our heritage **Jacob sheep** (F) help maintain plant diversity by discouraging unwanted species in our orchards and cycling nutrients. My root knife (or Hori Hori) is required in a wild landscape to harvest fresh dandelion and echinacea roots for a **spring tea** (G), which I am sipping as I write this, contemplating *life as tools* for garden success and life success.

Life as Tools

Pro-Tip: *Learn to recognize plants, soil, and animals as tools (and allies), as they have been regarded for millennia. Yes, we now use larger tools, but it is important to understand the beauty of life as tools/equipment. With this in mind, we can venture into the modern realm of pro-grower tools for homes, gardens, and homesteads.*

Tools can be used for any job to make it easier and more efficient. I collect sweet acorns in urban areas to add genetics to the Edible Biodiversity Conservation Area research trials with the goal of enhancing native tree species in food forest garden systems—an ecosystem restoration development.

FOCUS

Coppicing for Tool Handles

I head into one of our **agro-forestry projects** (1), an intensive **coppice woodland** (2) that is cultivated to provide limbs that can be used as fence posts, bows, garden tool handles, and other useful tools. When the trees reach a **desired diamete**r (3), I cut them to a stump and allow the replacement growth to regenerate from **dormant buds** at the base of the stem or **adventitious buds** near the cut surface—to provide material for future uses. Weedy, fast-growing tree species are encouraged here. Sometimes, we find unique opportunities, like volunteer elms that we can leave in place to grow a **thicket** (4) along with the planted trees to provide small-diameter wood for various projects (these trees wouldn't ever get big due to Dutch elm disease). Every tree has a unique shape. After a while, you start to *see the tool you want* in the tree, like this naturally strong fork in an **elm** (5) for a rake, the curved end in this **boxelder** (6) for a hilling hoe, and the straight body of this **hard maple** (7) for a universal tool handle.

Traditional Tools

You can still find tools made the old way. Many are beautiful works to behold and can still be used on the farm or in the garden. The application of human power directly into the ground will always be relevant. Cutting grass, grain, and hay with a scythe or sickle is still an affordable and efficient means of small-plot management. Many of our most popular tools today are based on traditional designs. The trowel and root knife have their own roots in ancient tools like the **Hori Hori**, which has been part of Japanese culture and gardening for centuries. (A picture of one appears in "Life as Tools," above [G]). It is a multi-functional tool used as easily for troweling out holes for transplants as for cutting below-ground roots, dividing rhizomes, digging up weeds, and even cutting greens. The Hori Hori has been used not just by *farmers,* but also by *warriors*. Many swords were forbidden in the 1800s, so farmers and blacksmiths perfected the use and forging of a hidden warrior's tool. It is fitting that a serious gardener or small-scale farmer holds a Hori Hori or similar tool at their belt in an age where these growers are indeed the Earth warriors pushing the world forward to a better place.

TOOL TIME

Traditional Tools

We are familiar with traditional tools like the **hoe** (1) which has been used to work dry and wet soils since the start of agriculture. Let's go back further and remember the humble **guard dog** (2), the first domesticated animal—still used to keep our flocks and gardens safe from predators and scavengers. One of the most powerful and longest in-use tools out there is **fire** (3); it has been used for tens of thousands of years to clear land, change environments, and encourage food production and consumption. At one time, fire was one of the greatest allies in agriculture. This *swidden agriculture*, practiced around the world, used fire to clear land and release nutrients. The Milpa system is a great example of using fire to grow crops. Semi-nomadic peoples would plant the land into annual crops and perennial fruits. After a period of production, they would move on to a new plot, leaving behind a patchwork of land with a new diversity of forest including more food-bearing trees and habitat for wild game.

The tools to make fire include a hand drill with a **hearth board** (A) and **hand drill spindle** (B) that use friction to ignite **tinder,** such as **tree fungus** (C) or fuel **fibers** (D) from coconut, palm fiber, cedar, juniper, lichen, or dried corn husks. Using a **flint** (E) is another method to spark the fuel. A simple modern **fire starter tool kit** (F) can be a tin holding various natural and manmade tinders, like a **char cloth** (G), and **flint and steel** (H) for sparking.

Original Tools

All about Tools

Familiarity with the diversity of tools and their forms and functions can help us make better decisions for our gardens and farms. You are already familiar with the basic types: the rake, spade, fork, etc., but there are many specialized tools, like the push seeder and flame weeder, that are also worth knowing about. Before we explore the use of tools in Part 2 (as they are used chronologically through seasonal operations), let's first get more familiar with the tools themselves, their architecture and types, and how form and function go hand-in-hand.

Architecture of a Tool

The most common tool architecture can be seen in any hand tool used for working the earth—from digging, to hilling, to furrowing, weeding, etc. As an example of **tool architecture**, consider this *spade's* construction (illustrated at right). It's a good spade because it has a strong **blade** with a **sharp cutting edge** that can be honed when needed. The **step** has a good ergonomic design. It has a wider step that has grip (grooved surfaces), which makes a big difference for safe and enjoyable use. This tool has a very strong **collar** that is well attached to the **shaft** with good-quality fasteners. (The collar is the weakest link on many tools; it is where they break if poorly made or poorly maintained.) There is also a **joint** where the handle attaches to the shaft. Sometimes this joint is a collar connecting the shaft to the handle, and sometimes it is made of the same material and is of a piece with either the shaft or the handle. The **grip** is where you place your hands, and this is another feature that should have good ergonomics. A well-designed grip like the one shown here will greatly improve your garden enjoyment.

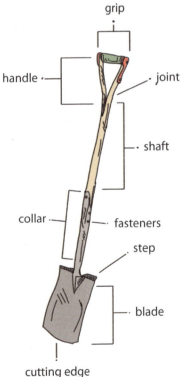

Tool Architecture

Collars and Fasteners

Collars and fasteners hold shafts to their various blades. The way this "business end" of the tool is attached to the handle is critical for longevity—this component is prone to break first. Although stronger collars resist breakage, certain designs lend themselves to being repaired more easily. If a blade breaks, it's a job for the forge or welder. But if the tool head is loose at the collar or the handle breaks, you'll appreciate having selected a tool with an easy-to-repair design.

This potting **soil scoop** (A) has a metal tang inserted into a molded plastic collar with no visible fasteners. It would only be strong enough for light-duty jobs; when any tool with this construction breaks, it's tedious to repair. This **pottery knife** (B) is assembled like most knives, with the metal tang (the solid forged metal part attached to the blade) inserted between two wooden handle pieces and fastened with rivets. This is a strong design for a handle; worn rivets can be removed, and new ones added. The metal collar for the first **hoe** (C) shown in the photo is fitted by forcing the hoe's metal tang into the wood, and expansion is limited by the metal collar. This is a popular design for metal files, but on a hoe, the metal will eventually come loose under heavy use. The handle on the other **pull hoe** (D) can more easily be fixed by simply removing a screw or bolt and inserting a new handle. The **edging tool** (E) has a variation of the metal collar on the first hoe (C), but it has a more substantial, longer tang inside. The small **hand spade** (F) has a metal insert collar that is much deeper than the **hoe** (D), and the wood handle is perfectly tapered for a snug fit. Considering that the long hoe will have more wear at the collar point (from pressure applied by a standing person whose hands are far from the blade compared to a short spade–handle distance), this spade collar is built very strong indeed. This **clevis hoe** (G) uses a standard split handle-and-wedge system to mount the blade. These eventually come loose, but it is easy to replace them by hammering a new wedge into the glued channel at the handle end (then break the wedge off and sand it clean). A second metal wedge can be hammered in across the first to further strengthen, but this metal wedge can be obnoxious to remove from an old hoe head! The multi-connection-style handle shown with the **furrower attachment** (H) on the right of the image fits many tool blades, but it has a completely custom collar, so repair requires ordering parts from the manufacturer.

TOOL VIEW: Collars and Fasteners

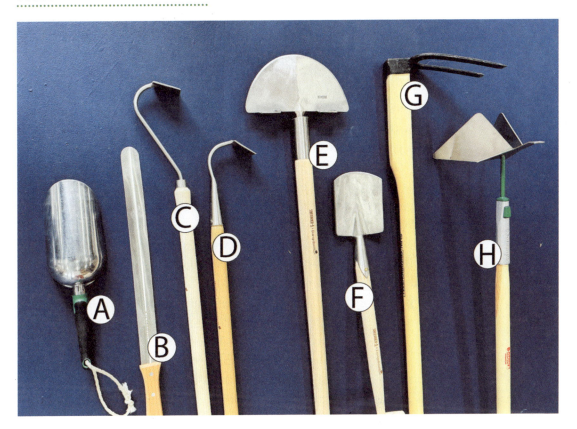

Handles, Joints, and Grips

When it comes to ergonomics, hand grip is critical (along with having the proper shaft length to suit your job and height). We can also look at the different joints of the handle for strength considerations.

The **joint** pictured on the left of the image (I) is tapering slightly to the handle, whereas the **shaft joint** (J) is straight into the handle grip. The Y-joint (K) is made of a solid piece of wood, whereas the Y-joint (L) has a **triangular support** of wood blocked between the divided handle wood coming out of the shaft of the tool. The tool (M) has a **metal Y-joint** to join the shaft and the grip, and the tool (N) has a **plastic** shaft and joint. The strongest of these are the **solid wooden Y-joint** (K) and the high-quality **molded plastic handle** (N). The most difficult one to repair would be the plastic handle, followed by the split wooden one with a triangular block (L). The most likely point of failure will be on the model (M), with its **metal joint** that is likely to become loose with time (actually, this one is!). This is the only model where a broken shaft doesn't require replacing the joint and the handle grip.

For grips, note that some designs (P, S, T) **taper at the edges**, whereas others (Q, R, U) **taper upward** at the center of the grip. These will feel different in different hands. Overall the **T-handle** style (P, Q) is better for tools for digging into hard earth (spades, digging forks, edgers) with your primary hand on the grip, and the **D-handle** style (R, S) is better for when the handle is held with the primary hand and the second hand holds the shaft for moving materials around (shovels, garden forks, hay forks, etc.). *Note:* This is also true for tools (like a fork) that are primarily for lifting material like soil and moving it up and into a space (like a wheelbarrow) or turning it over (mixing compost). On the other hand, long tools with a **straight shaft/handle** (O, V) are best for when you need longer reach with the tool and/or when the action is more downward. A long-handled shovel, for example, is used to skim the soil surface and turn the lightly moved material onto a row cover to hold it down. One of my favorite shovels had a pointed and curved blade to cut into soil and scoop and hold material; it was attached to a long, straight handle for easy reach. A straight handle will be found on earthwork hoes for *picking* into new ground (an action where a D- or T-handle would never be used) because hand positioning requires the straight shaft. The most common is the variety of garden weeding hoes with a straight handle that the user holds upright with both thumbs up so the hoe can ergonomically be used to work rows of crops.

There are variations in grip *material* as well: **metal** (O) is cold and not fun in rain, but it's strong; a **foam** outer layer (V), which is soft and warm; a textured **rubber** grip (W), which is anti-slip and warm to the touch; and **wood** (X), which is also warm to the touch and anti-slip when finished correctly. Another option is the **stacked leather** grip on this **billhook** (Y), which is very cozy to hold. The classic **hard plastic** grip on this **harvest knife** (Z) is easy to keep clean, so it is very good for the fine work of pre-washed greens.

TOOL VIEW: Handles, Joints, and Grips

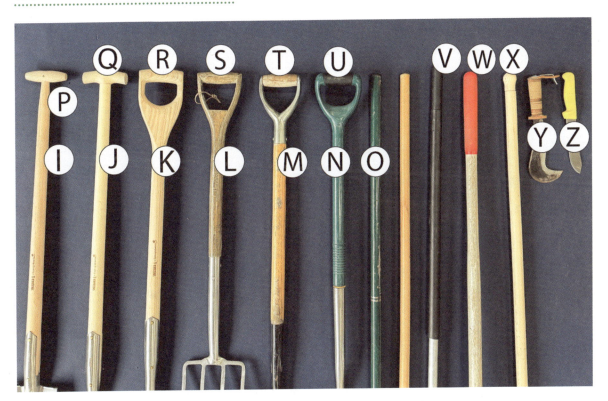

Specialized Tool Components

Most tools have similar components, but there are some differences. The **wheel hoe** (1) is noticeably different from your typical **hand tool** (2) because it (obviously) has wheels; the version shown here is a double-wheel hoe with a special toolbar for attachments. The attachment here is a tine weeder and torsion weeder, not unlike the tool head on the hand tool (2), which is a spring hoe. The **weed wacker** (3) is also different because it has an electric battery power source, but otherwise, it is essentially a scythe. Smaller hand tools like the **sledge mallet** (4) have handles and heads, and the **knife** (5) has a handle and a blade not unlike the spring hoe (handle and blade), but in this case, the blade of the knife is much sharper and the handle much shorter. The pruning scissors and **clippers** (6) have similar components as well, but their form has a different intention. The **backpack sprayer** (7) has straps for your back (similar to those on the weed wacker), but it adds several new components: a tank for solutions, a piston, a hand pump, and a wand with a sprayer nozzle. The **push seeder** (8) is also very different. It has wheels that drive a belt that operates the turning seed plate in the hopper. Some simple tools, like the **ripper** (9), are very specialized. Most tools are some version of these elements, with handles, blades, wheels, batteries, or hand-operated mechanisms. In all cases, the form of the tool is based on its function.

Specialized Tool Components

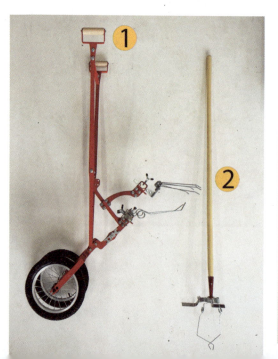

All about Tools   **17**

Tool Types

All tools come in many shapes and sizes, from the simplest dibble to specialized tractors. Most garden tools can be categorized as hand tools (meaning you hold them in your hands), but this is vague. Does that include hoes, power drills, push seeders, and garden carts, but not salad spinners? Garden tools are best categorized by the type name: shovel, seeder, trowel, fork, etc., and they are best understood by the jobs they perform. The categories overlap: a long-handled tool can be multi-functional and multi-row. However, the categories given below help to define the basic aspects of tools we should consider when creating complete tool systems.

Tool Type Categories

1. **Short-handle tools,** like trowels, have handles around 6" to 9" long and require human proximity to what they are doing. You use them when kneeling in the soil to transplant or standing at the edge of a container bed to plant. These are popular for container growing, pot culture, and greenhouse work.
2. **Long-handle tools,** like hilling hoes, allow you to stand and walk while working the ground. These are popular with market growers and anyone working in raised beds between 2" and 12" in height.
3. **Medium-handle tools,** specifically designed for container growing, have handles long enough to work the soil in raised container beds that are 12"–30". Medium-handle tools like picks and broad hoes are also employed in hilly terrain where you stand downslope—you don't need the more typical long handle to break the ground uphill.
4. **Wheel tools,** like push seeders, wheel hoes, and carts, have the momentum of wheels; some make use of ground-driven power, as a push seeder does, with the front wheel driving a belt that turns the seed plate and plucks seed from the hopper to deposit them into the ground through a furrower.
5. **Multi-functional tools** are simply those that have many uses. A garden fork can be used to turn over soil, harvest root crops, and move compost into a wheelbarrow. You can find more specialized tools for such garden tasks, but multi-functional tools are invaluable, especially on

a small farm with many tasks that need to be done in a cost-effective way.

6. **Dual-purpose tools** include those that have been intentionally designed as two different tools in one, like the pickaxe, which is literally a pick on one side for breaking hard soil and an axe on the other side for chopping through roots when opening new land.
7. **Attachment tool systems** use a base tool that can be used with several attachments to serve different tasks. Long-handle tools can have multiple hoe heads that are interchangeable, while wheel hoes can use a system of different belly- and rear-mounted tool bars to attach different hoe types as well as other tool types like cultivators, hillers, and tine weeders.
8. **Multi-row tools** are popular with larger-scale growers because they perform the same job (weeding or seeding) across an entire bed top. For example, the 6-row seeder or tine harrow attachment for a wheel hoe can be used to seed an entire bed top and then blind weed the whole bed top.
9. **Power tools** are those with power (whether gas or electric); they can be pushed wheel tools, short handheld, or long handheld tools.
10. **Stationary tools** are those that have a fixed position and perform a task, like a root washer or salad spinner.
11. **Bins, Containers, and Trays** are all those vessels that hold what we need—from the seedling tray to the produce bin to the raised bed made of wood or metal.
12. **Irrigation** is really its own category because it is a *system* that only works when all its components are working together.
13. **DIY tools** are those built by growers to suit their needs; they might be innovatively purpose-built or built using design templates.
14. **Supplies** are anything that has a shelf life, including weed barrier, drip tape, and row cover. Proper handling of these will get you maximum longevity.
15. **Inputs** include immediate consumables like fertilizers, compost, and pest products.
16. **Life (ecosystem services)** includes the living organisms that serve your garden production—from soil microbiology to manure producers to the people who help it all run.

Tools Systems

Most tools work together to form a tool system—accomplishing a series of tasks. In some cases, the *system* is clearly a system, like an irrigation system, with its various components that pump and distribute water to your crops. A wheel hoe, though, can be seen as either a singular tool or, if various attachments are used, it could be seen as a system that maximizes humans' upper body strength and the quick movement of the wheel to accomplish different tasks. What type of weeding system do you use? We use a *wheel hoe tool system* for our garden cash crops, like carrots.

Hand trowels are short-handle tools; push seeders are wheel tools; chainsaws are power tools; and irrigation is best understood as a system.

Tool Types   21

FOCUS
Attachment Tool Systems

For the home gardener, a practically placed tool wall with well-built, forged tools for each garden task will probably always be a mainstay. Yet, increasingly, tool design is shifting toward *multi-functional interchangeable tool heads* on the **same handle** (A) to achieve different garden tasks. Johnny's Selected Seed's *Connecta* system is an example of this; it has some of the most commonly used tools for key garden tasks such as fine seedbed prep with a **rake** (B), row marking the **matrix gridding system** head (C), and **two-row** (D), **four-row** (E) **and three-row** (F) markers; furrowing and hilling with the **Row Pro** attachment (G); and cultivation with **collinear hoes** (H) and various sized **wire weeders** (I). This is an example of the systems thinking that takes a grower from start-up to pro-up, namely, having the full suite of tools necessary to do a good job. I look forward to seeing more tools in this category, as it is very practical for landscapers needing to fit many tools in a trailer. Similarly, it works for community growers who store tools in small, locked sheds on site. It is also useful for the market gardener who will enjoy throwing 3–5 tool heads with one handle into a cart and heading out to the field. For myself, as an edible ecosystem designer/landscaper and nomadic gardener, I find it handy to carry consolidated tools with me to build edible research sites, called EPI (Education, Propagation, Inspiration) sites, across the continent. I build these sites and return to them to harvest, research, and teach, so tools that break down into a duffel bag are great!

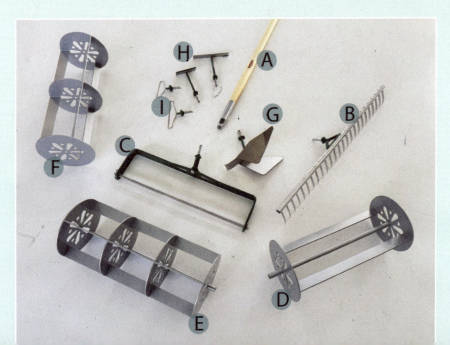

Connect a System

Tool Design

There are so many different tool variations out there. The variety of blade shapes on hoes or tines on rakes is staggering. It's the same with spades and shovels as well as task-specific hand tools. Each tells a story of form for function.

Form for Function of Tools

Understanding the nuances in tool design is part of what separates start-up growers from pro growers. Small design shifts make all the difference for garden-task efficiency; sometimes they are relevant to your soil conditions, and sometimes to your management style for specific crops. For instance, the typical garden hoe with a **broad blade** is ideal for hilling. A variation with a narrow blade, called a **collinear hoe**, is great for cutting tender weeds. But you may want the **stirrup hoe**, with its oscillating function and an opening between the handle and the blade; its oscillating action is well suited to stony soil or bed tops with lots of garden debris (crop residue).

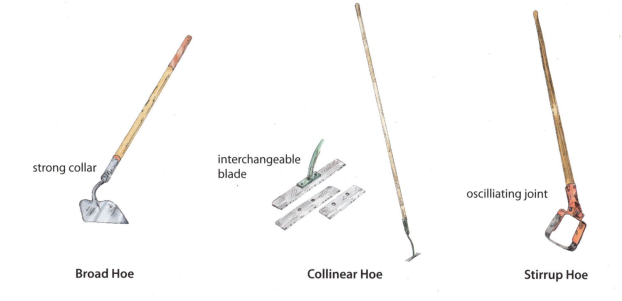

Broad Hoe · Collinear Hoe · Stirrup Hoe

Hoes

Tools come in families, but they don't necessarily share the same name. For instance, the **wire weeders** (I and H) and the **furrower** (G) are very different from the rest of the hoes pictured here, like the **hilling hoe** (F) and the **Dutch push hoe** (D). But ultimately, these are all hoes—designed to weed by moving soil between and around crops.

Another way of thinking about tool families is their form. As an example, the *tine family* of tools (see photo just below, under "Rakes") which includes **rock rakes** (R), **leaf rakes** (Q), **scarifying rakes** (N), and **cultivators** (K), are task-specific due to variations in the length, thickness, strength, quantity, and width of their tines. What is important is understanding how tool shape serves our purpose: are the functions of one tool adaptable to a different task? As an example, I used a normal garden rake to stale seedbed (pre-weeding a garden bed before seeding) for years before getting a tine weeder.

When it comes to holding my long-handle hoe, I always remember what I learned from Eliot Coleman: both thumbs go up!

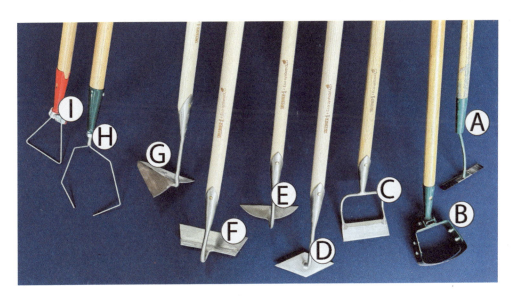

TOOL VIEW: Hoes

A. collinear hoe, B. stirrup hoe, C. Dutch push hoe, D. diamond push hoe, E. halfmoon pull hoe, F. pull or hilling hoe, G. furrower, H. wire weeder (interchangeable model has other wire head sizes), I. delta wire weeder

Rakes

Consider **rakes** as another example of the variation in tool designs and how different models relate in their form to their function. The typical narrow **garden rake** (L) is great for smaller bed tops and paths but will take far too many passes to even out a 32" bed top; for this, the wider garden **preparation rake** (M) is far superior. The **rock rake** (R), with its strong, short tines fixed at a 90-angle from the rake body and handle, is best for pulling through heavy material because the tines won't bend or break.

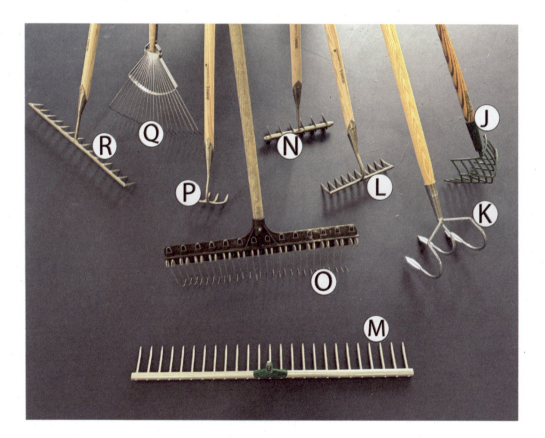

TOOL VIEW: Rakes

J. rock rake, K. 3-tooth cultivator, L. garden rake, M. wide garden rake (Connecta model), N. scarifying rake, O. tine weeder, P. rose/weeding rake, Q. leaf rake, R. rock rake

Shovels and Spades

Consider the **shovel** and the **spade**, tools that are often confused, misnamed—and misused. A shovel is for scooping material, and a spade is for breaking into hard soils. But variations and hybrids exist for doing particular tasks. This **Celtic shovel** (A) has a collar bent at the blade-to-handle joint, allowing for a very ergonomic motion while standing and moving material from a flat surface—for example, when skimming horizontal paths to level them off. The **transplant spade** (D) is meant for surgical working around perennial shrubs to dig them up and to plant them in. It does this much better than a **border spade** (G), for example. Yet, form is function, and the **stone spade** (E) will do this job better in stony ground. The **drainage spade** (C) cuts a narrow trench for creating drain fields or laying waterlines.

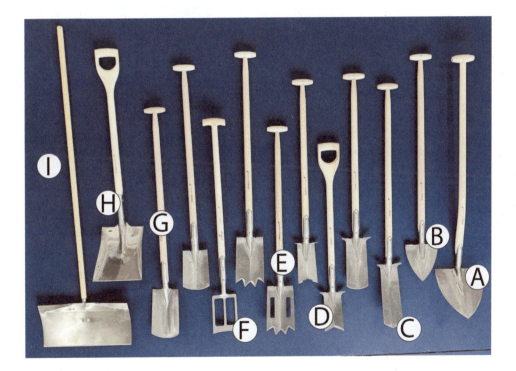

TOOL VIEW: Shovels and Spades

A. Celtic shovel/spade, B. pointed spade, C. draining spade, D. transplanting spade, E. stone spade with steps and openings, F. field shef, G. spade with step, H. square shovel, I. snow shovel

Task-Specific Tools

Let's also consider the form and function of some task-specific tools, their designs, and the functions they serve. The **dandelion digger** (J) is formed to lift these tough roots, whereas the **weeding fork** (X) works better for loosening soil around fibrous-rooted weeds to pull them out. In another realm is the flat **stone scratcher** (U) that can weed between paving or patio stones or around the edges of beds.

TOOL VIEW: Task-Specific Tools

J. dandelion digger, K. hand scratcher, L. hand hoe, M. Von Lindern cultivator, N. hand cultivator, O. pull hoe and fork, P. wooden dibbler, Q. pottery/container knife, R. hand (chisel) cultivator, S. bulb planter, T. finger weeder, U. flat stone scratcher, V. royal Dutch hand hoe, W. dandelion digger, X. weeding fork, Y. wrotter, Z. asparagus knife

Task-Specific Knives

The best way to understand how to grow great garden vegetables is to understand the seasonal tasks that must be performed for each crop. Some tasks are the same for multiple crops, while other vegetables have tasks very specific to their nature and growth. For instance, most crops require the garden to be prepared for planting or seeding, but the preparation for transplanting is different than for seeding. Additionally, some crops (like leeks) need to be hilled, while others (like broccoli) don't. Consider the job at hand and the nuance needed for certain crops to guide you to the right tool.

Understanding the multi-functional utility of a tool also means you understand how to use the tool properly when doing a task that the tool isn't specifically designed for. You can see how a rake can be used to pre-weed by killing small germinating weeds in a bed, but you wouldn't use a rake to wack at weeds or hill your potatoes (wrong tool for the job).

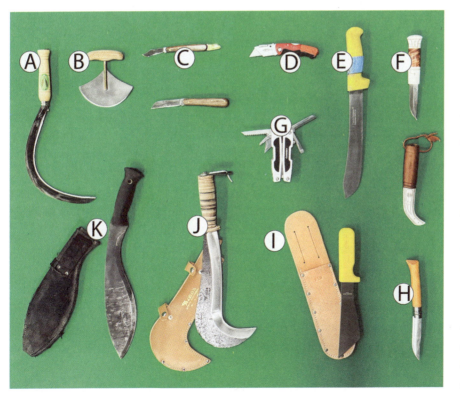

TOOL VIEW:

Task-Specific Knives

A knife may be a knife, but when it comes to tasks in your garden there are many specific types. Here are some examples: A. hand sickle, B. ulu, C. grafting knife, D. box cutter, E. harvest machete, F. Sami knife, G. multi-tool, H. folding pocket knife, I. lettuce field knife, J. billhook, K. woodland machete

Trowel Heritage

The many varieties of trowel that exist each tell a story of form and function, but also the regional heritage and individuality of gardeners. Trowels are primarily designed based on how they need to make holes in the soil for planting. Some trowels dig a hole by inserting the trowel and scooping out soil. Others are meant to be inserted and pulled horizontally to press the soil away from the slit opening for inserting a transplant. In other cases, the trowel can carve a hole through compacted sediment in the soil and even cut through roots and displace stones. Sometimes, a trowel is meant for making small openings for small plants and sometimes for bigger openings for bigger plants. If a transplant is really big, you will need to switch to a spade. A trowel is, in essence, a small spade called by another name.

Consider the **Christopher Lloyd trowel** (A), named after the well-known English country gardener and author. It's a great, deep digger for narrow plug plants. The **Dutch-style planting trowel** (B) has its roots in the Dutch bulb industry and is used every year by the Dutch Keukenhof Exhibition gardeners to plant millions of bulbs in their gardens. Note the broader blade that makes the shallow but wide hole you need for tulips, garlic, or other bulbs. Its longer handle lends itself to the fast, efficient motions of bulk planting. A variation on this bulb-oriented trowel design has a romantic twist—the **heart-shaped trowel** (C) has a smaller size and shape that can be ideal for home gardeners. The **container trowel** (D) has a unique flat blade that is ideal for opening large holes in container planters. The design is not conducive to digging and lifting material; it is meant to slice into the soil and be worked back and forth to open a wide slit—useful in loose pot soils where there is nowhere to put the dug-out material (better to displace it than remove it). The flat blade is also useful for digging and scraping against sides when working at the edge of a container or pot.

The **potting trowel** (E/F) is designed to fill containers and pots with new soil material. Its unique curving blade makes it easy to fill round and/or tight spaces without spilling material over the edges. This model from Sneeboer & Zn has a **right-handed** (E) version that allows the user to twist it clockwise, and a **left-handed version** (F) for counter-clockwise digging. A build that is adapted to the user's hand further improves the agility of the gardener. A similar round shape can be seen in the **flowerbed trowel** (G);

Tool Design   **29**

its curved and tapered shape allows you to transplant new plants between existing plants in flower beds without damaging them. This design is also good for lifting plants out by scooping around them to transplant into a new container. The **weeding trowel** (H) is specifically designed to maximize the removal of weeds with its narrow, long V-shape. This strong trowel is good for prying out roots like burdocks and dandelions.

So, why have a weeding trowel versus some other weeding tool? Because they are multi-functional. Weeding trowels can be used to remove weeds, plant transplants, and even harvest root crops, like perennial echinacea root.

Sometimes it is the soil conditions that dictate trowel design, as is the case for this **archeology trowel** (I) that can easily slice into hard and compacted

TOOL VIEW:
Trowel Heritage

soil layers and clay—carving the planting hole rather than digging it out or opening it up, as other trowels do. I find this trowel useful when adding perennial herbs into existing fruit tree rows where the soil is packed and not tilled. If you need to take this action to the next level for planting or digging out plants in hard soil, the **Tissot trowel** (J) offers the precision of a short-handle tool and maximum blade sharpness and durability. The V-notched blade can cut through small roots and displace stones.

One of my favorite designs is the **Kappe trowel** (K); it has a very multi-functional blade that is strong and wide and has a good curved shape, making this trowel a nice middle ground for digging out holes, filling pots, and slicing through roots.

For those working quickly in prepared garden beds with many transplants, consider the **Helmantel trowel** (L). This will be familiar to market gardeners used to planting long rows of hundreds of transplants into loose, prepared beds. The 90-degree angle of the blade and handle allows the user to pick it into the soil easily and pull it back to open an efficient space for inserting a transplant.

The **greenhouse trowel** (M) is a very small-bladed trowel used for planting and transplanting young plants into loose material where hardly any power is needed to make holes. It can also be used between plants or in small pots. Trowels can be quirky too, like this transplanting **trowel with bottle opener** (N) that is handy on a hot summer afternoon.

When considering ergonomics, we can also look at the **Arends trowel** (O). Its angle at the handle and blade attachment joint reduces user fatigue, and its shape is very effective for opening holes and scooping material. This trowel might look familiar—it is one of the most popular shapes out there. But, as we've just seen, specialized trowels can be a means of increased efficiency (and pleasure!).

A classic trowel design is seen in the **half-round transplanting trowel** (P), with a form meant to make deep holes for transplanting vegetable starts. Its sharp blade cuts through roots and clay layers with ease, and it has a concave blade that allows you to easily scoop out the holes for placing plants. A fun variation I like for packing out on research excursions is the **multi-shovel** (Q), which is essentially a little trowel plus a cultivator. It would also be practical for small pot production—allowing a bit of both digging and cultivation to be done in one hand with a twirl of the baton.

Forking Forks

I will admit, my favorite garden tool is the fork. Maybe it is because I need them so often. I used to grow thousands of carrots and needed sturdy digging forks to lift them from my heavy clay soil—and the broadforks were essential to breaking up that heavy clay soil to plant them initially! Or perhaps it is the 100+ varieties of garlic I grow that need a good border fork to lift them and hay forks to mulch them in for winter.

Forks have very different uses despite their similar names. Form follows function for the fork. (I know. That's a lot of alliteration!) They fall into a few broad categories. **Broadforks** have many long and strong tines for deep digging. **Garden (English) forks** are heavy-duty and often have more curved and rounder tines, allowing for multiple uses, including turning soil and compost and moving debris during fall cleanup. But they don't work as well for digging root crops because the sharp and curved tines damage root vegetables; their sharp tines are meant for cutting through roots when preparing a new garden bed—not your perfect vegetables! They are generally used to loosen soil in the garden before a new year's planting. A smaller garden fork is called a **border fork**—which can be used more like a spade. They are a form of **digging (spading) fork**, which generally have four shorter, straighter, flat-faced, and blunt-tipped tines. This makes them good for digging deeply in the garden bed (penetrating the soil layers is easier with straight tines), allowing aeration and double-digging and garden bed turnover. They are good for harvesting root vegetables—their straight, blunt tines won't harm the roots.

To help remember the major differences among forks, focus on the tine form for the function. The *curve-tined* **garden fork** is more like a shovel, used for moving material like compost and loose soil; its sharp tines stick through strawy compost, but they will harm your root crops. The *straight-tined* **digging fork** is more like a spade for breaking into the soil to aerate new garden beds, and its dull-tipped tines won't damage your carrots, potatoes, and garlic.

Hay forks have fewer, narrow tines for lifting straw and hay. Whereas **chip mulch forks** and **compost forks** have many long tines for lifting loads of material—the former with knob-ended tines to hold material like a shovel and the latter with sharp, hay-style pointed tines to stab through wet composted manure with strawy material in it. (**Note:** An actual shovel is useful for very loose finished compost.) Most of the rest of the variation in forks is in handle length or design for specific purposes, like container gardening or rose growing.

Fork Models

A **broadfork** (A) can have round tines or flat; the latter is best for breaking up old garden bed hardpan, creating deep aeration. The **6-tine digging fork** (B) can also turn over old beds and is excellent for harvesting long rows of carrots and similar root vegetables. The classic **English garden fork** (C) is a great multi-functional tool useful for turning soil and moving debris. The **4-tine digging fork** (D) is much narrower than the 6-tine version, allowing it to be used in narrower spaces and more easily push deep into the soil to aerate beds.

The variation in tine length and handle length has to do with user preference and bed types. Shorter handles are used for container growing and longer tines for deeper turning of the soil. The smaller models shown on the bottom left of the photograph are "border forks" because they can fit into the tighter borders around garden beds. (Sometimes they are referred to as "lady forks" because they are lighter and easier to use. But this may be a bit of anti-feminism: my wife likes the bigger digging forks for turning her garden beds, and I like hauling around lightweight but strong-tined digging forks for odd jobs in the food forest—so the border fork suits me perfectly!) **Three-tine *digging* forks** (E) are shy one tine, making them even easier to work into dense or root-crowded soils for double-digging and aeration. Whereas the **3-tine *garden* fork** (F) has a curved tine better suited for moving material and turning compost. The short-tine model shown here is ideal for lifting small perennials like strawberry runners for transplanting. This **ground elder fork** (G) has an aggressively concave set of four tines that end in sharp points to help lift shallow-rooted plants like ground elder (aka Bishop's weed), and other spreading weeds out of the soil. The **2-tine rose fork** (H) is used to loosen soil around roses and other perennials. Its two tines slip easily around perennial roots to open soil and improve drainage and airflow. The **3-tine pitchfork** (I and J) is designed to move straw onto beds for mulching and feed hay to animals. If you are

Tool Design   33

working with small or large square bales, you will have smaller or larger forks to handle the broken pieces of the bale.

A **gravel/chip mulch fork** (K) is used to move wood chips and gravel with a scooping action; this fork can hold material while still being able to *stab* into piles in a way a shovel can't. A **compost fork** (L) is designed with sharp, pointed tines to pierce straw and hay-like compost (composted manure) as well as chips because those tines are also close together, forming a shovel-like shape. This **chip mulch fork** (M) is great for moving and holding woody material. The **"The Bear" digging fork** (N) is a beast, able to move a lot of material and dig deeply. This chimera of a **compost fork** (O) is a hybrid between the garden fork and the compost fork, allowing turning of the earth easily with only four tines, like a garden fork, but much sharper and flaying out at the end. It won't turn packed soil, but it works nicely to turn loose compost into loose soil, move spent garden crop debris, and even mulch with hay and move composted manure about. This **3-tine potato fork** (P) is ideal for potato harvesting and helpful for removing weedy perennials by sorting the roots out of the soil and lifting the whole plant in its scoop-like head. This **Great Dixter tickling fork** (Q), designed

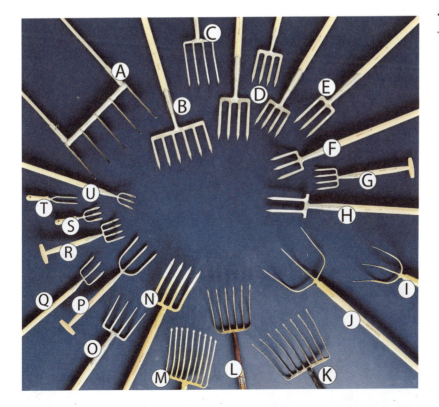

TOOL VIEW: Fork Models

by Christopher Lloyd, is a practical tool suited to the dense self-sowing perennial Eden at his Sussex home, where getting at the weeds amidst the shrubs and tall herbs required a long-handled, small-tined fork. The **Great Dixter fork** (R) is designed as a spading fork for working in a kneeling position in flower and other garden beds when transplanting perennials. It is similar to the ground elder fork, with its longer handle for more leverage with deeper-rooted plants. The classic **weeding fork** (S) is the short-handled fork for weeding and harvesting small root crops like radishes in container beds. The **greenhouse weeding fork** (T) is a 2-tine, short-handled fork for loosening weeds around dense plantings in the tight spaces of a greenhouse. This **weeding fork** (U) is another long-handled version of a small handheld weeding fork; it's used to get in between perennials to dislodge weeds.

TOOL TIME
Broadforks

The **broadfork** is a large digging fork that allows deep penetration; tine lengths are greater than digging and garden forks (usually 12"–16"), with a slight taper to help the operator lift the soil. You place the tines into the soil with the handle slightly forward, step on the bar, and press down firmly; you then rock the fork back with your hands, with your feet on solid ground—lifting the soil. There are three general types: the **cracking fork** (1) has a metal bar rack that stops the lifted soil and cracks it apart; this works well in loam soils and prevents the broadforking from leaving an uneven bed surface. The **hardpan fork** (2) has flat, rectangular, tapered metal tines that are very strong and can penetrate clay and break it apart, and it can deal with stony soil—all without bending. The **traditional broadfork model** (3) had narrow tines that penetrate soil with ease; it works great in already-prepared beds without heavy soil layers or lots of stones. This model will take less work to broadfork a suitable soil than the heavier hardpan fork.

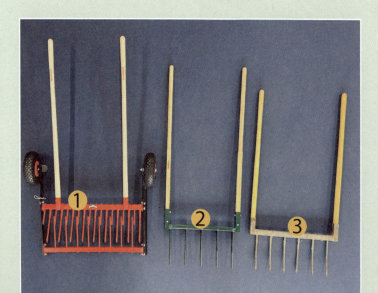

DIY or Buy

There is great benefit to buying a well-made tool, but there is also great joy and practicality in making one yourself—or creating a design and having someone make it for you. Each option has its own cost, time, and energy to balance. If you like to tinker, building your own is a sure bet; if you have a green thumb, but aren't so handy with a skill saw, then buying most tools makes more sense. Sometimes we build for savings and sometimes to fine-tune our operations. Always, what is most important is getting the right tool for the job, whether you DIY or buy.

Scaling-Up with DIY Tools

Building and innovating are a huge part of gardening, homesteading, and small-scale farming. As growers, we often need to solve *weak links* in our production with custom design/builds. As we **scale-up** our garden operations, we will always encounter these weak links in our operation (see more on scaling-up later). DIY projects offer a low-risk means of scaling-up your operation. For example, before investing in low tunnels, we build cold frames; before we invest in high tunnels, we bend our own PVC hoops. Many growers I have spoken with have first made their own solutions to tool and infrastructure problems before committing to large purchases. In some cases, the DIY tool is all that is ever needed.

Budgeting for DIY

Do you always save with a DIY project? In truth, DIY can cost *more*—if your design isn't right or you don't truly understand what is needed to fix the weak links in your garden operation. In other cases, it may cost less, but maybe the quality isn't good enough. Often, though, a DIY project is a perfect blend of the cost-effective and the functional. Sometimes a DIY project is as simple as repurposing a found object for a new, practical solution.

The drip roller featured in the photo below cost $70 in materials and time. Some growers have used their cart and rebar to make a drip roller—for the cost of a 4' piece of rebar. There aren't many drip tape rollers on the market.

Hose rollers are more common. These are often plastic, poorly constructed, and low to the ground, requiring bending over. So the trend in drip tape rollers being DIY-built on small farms is partly out of necessity.

FARM FEATURE
DIY Drip Roller at Cully Neighborhood Farm

One of the most useful tools to build yourself is a handy drip tape roller. It can help market growers scale-up through the reuse of drip tape. The design seen in the photo was found at the **Cully Neighborhood Farm** (1) in Portland (and it wasn't the only great DIY tool project I had the pleasure of viewing the day I visited the farm). This design is composed of a **large spool** (2) that holds the **drip tape** (3). The base has a **cradle** (4) supported by a four-legged braced **stand** (5) where the spool is attached using a galvanized **flange** (6) and a **threaded rod** (7) set into a notch in the cradle. The other side has another flange and rod resting in a similar **notch** (8), but on this end, a series of elbows and threaded rod make a **handle** (9) for turning the entire spindle and winding up the drip tape. This design puts the grower in an ergonomic position to easily roll up drip tape while standing.

DIY or Buy 37

Josh Volk is a DIY tool guru. Here, he shows his small-scale wash station and explains the multi-functional cart designed to haul everything (note its wide easy on/off platform).

Simple, Cost-Effective DIY Projects

The photo here shows some cost-effective and simple DIY projects. They are the kind of easy-to-build projects that I really like; they serve their purpose perfectly and would be much more costly to buy: (1) **oak trellis stakes** and plant row label stakes for perennials were custom designed and ordered from local saw mill, then painted on the farm; (2) a **built-in cedar wash table** supports bagel trays for spraying bunches of carrots and other roots; (3) **bulk harvest bins** made from plywood on salvaged pallets; (4) stackable bulk **garlic curing trays** made from 1" pine boards; (5) reused bagel trays used as sorting/delivery trays for getting tomato quarts to market; (6) empty Rubbermaid **bins with burlap** on top for market shelving once the product is arranged on table; (7) Rubbermaid bins on harvest crates used as **giant ice cube trays** with drip spaghetti tubes filling them with water to make ice for cold storage; (8) root cellar **sorting table with pull-away slats** to drop "B-grade" and "C-grade" carrots into bins below so "A-grade" carrots can be bagged for sale; (9) blossom of **propagation guilds of perennial fruits** in trial plantings that are near-to-home for future far-out-to-field orchard plots—*life as tool concept* to a T.

DIY or Buy 39

Rolling Row Marker Tool from Scratch

Building your own tools can be a satisfying process, not only for the romance of working with metal and wood but also because you can bring flourish to designs that are truly fine-tuned to your operation. You can measure (1) and cut (2) handles to perfectly to suit your height. Sand them smooth (3) and finish them (4) with your preferred oils or paints. So, make a design, then line up (5), measure (6), mark (7), and get to work (8) making a custom tool for your needs. This rolling bed marker (9) is perfectly set up for the 123 Planting system (covered later in this book).

This rolling row marker (or as I call it, the "Hortigrapher," for horticultural graph paper) leaves a square grid across my bed tops, marks perfect 5" spacing between seven equidistant rows in one pass to help line up seeding. But it also marks cross lines at 5" spacing, leaving a crossline layout to help organize in-row spacing for transplanting. I built this tool to allow best row marking for my 123 Planting Method. In this method, the focus is on simplifying intensive multi-row plantings on Permabeds for gardening, market growing, and Permaculture polyculture plantings. Because this tool marks the entire bed top in graph-paper grid, you can design your plantings into the spacings; growers can simply seed or transplant the relevant rows at either 1, 2, 3, 4, 5, or 7 rows per bed and choose different in-row spacing too. Whereas other row-marking systems require adjustment of row-marking tubes, or wheels or dibbles, or different roller heads, this tool marks all beds the same, making it efficient to mark a whole plot and then plant and seed as desired. I find this to be a vastly simpler and more successful model for intensive vegetable growing in row-based systems on Permabeds or flat ground and much more conducive to designing polyculture plantings of annuals or perennials where some rows can be planted in early-maturing crops (for example, the middle row in peas and outer rows in longer maturing crops like beans).

Right: *This rolling row marker was custom made to suit my planting system of up to 7 rows at 5" apart on a wider 40" bed top.*

Below: *This rolling row marker can also be used on a more standard 32" bed top as pictured here, and the outer-most rolling discs run along the shoulder edges to help keep rows lined up on the bed top.*

FOCUS
DIY Projects

Here are some of the innovative DIY projects I have done over the years on different scales for different garden tasks: (1) **mobile analysis work station** is a stacked drawer system that houses metal plant name tags and is attached to a dolly for easy moving into the orchards, and my laptop comes out to rest on the desk-like top for data note taking in the field; (2) **sunroom potting and grow system** for suburban homestead, (3) bathtub **transplant-tray filling station** for market garden, (4) built-in **multi-functional tables** with ceiling- and wall-mounted water supply for hardening transplants and using as wash tables for vegetables, (5) water wheel transplanter system transformed into a **3-point hitch dibbler** for small-scale garlic farm, (6) multi-connector for irrigation quick connect in large fields, (7) **roof catchment** into 250 gal water tanks on a pallet for homestead irrigation, (8) **moveable hoop house** with overhead watering system, (9) PVC conduit bent for hoops over Permabed crops for season extension.

FARM FEATURE
Gemüsezeit Altluneberg GbR

Gemüsezeit Altluneberg GbR is a farm in Germany run by partners Sarah and Phillip Puckhaber. They sell directly to customers from their 3,230 ft² (300 m²) garden and 5,600 ft² (520 m²) of greenhouses. They are in year 5 of production, having scaled-up from hobby gardening. They have most of the tools they need: broadfork, wire weeder, Jang seeder, wheel hoe, tine harrow, greens harvester, and their DIY leek puncher. Their next big project is getting a cold storage facility.

Sarah does her own starts in the **greenhouse** (1), which has a rigid panel structure (A) made of long-lasting glass and metal as opposed to roll-out poly covers or polycarbonate panels. The plants are grown in **cell trays** (B). Phillip is **harvesting tomatoes** (2) in the high tunnel. Favorite **tomato varieties** (C) include Tica (pictured), Ruthje, Green Zebra, and Black Cherry. Their absolute favorite variety is Ruthje—locally bred, great storage, and amazing taste! Similar to many humid and cold-climate growers, the Puckhabers have taken to only growing tomatoes in the greenhouse to avoid fungal disease and increase the harvest. The tomatoes go into ridged **poly crates** (D) on a pull cart that goes along the alley. Greens grow along the edge (E). Efficient production requires the right tools and good timing. Here (3), Sarah uses the **flame weeder** (F) with propane in a **backpack** (G) to pre-weed the greenhouse beds. Healthy tomato transplants (4) are grown in 3" inch (7.5 cm) **pots** (H) until they are the right size for transplanting. Greenhouse **beds are irrigated** (5, I) using a **header hose** (J) and long-lasting heavy-duty **drip tape** (K). The outdoor **garden beds** (6, L) were initially prepared using a tiller, but now are simply broadforked and raked. Their **DIY dibbler** (M) is ready for imprinting the bed for planting, while the **drip kit** (N) is nearby to water them in. The dibbler is actually a dibbler *plus* a **leek puncher** (7) with longer **wooden punches** (6"/15 cm, P) than those on other dibblers, and the **dibble bars** (Q) help set the next row for punching. The operator presses them into the deeply worked bed, and the **tool handles** (O) stabilize and lift the dibbles out of the holes for the next row. The deeply **dibbled bed** (8) is **ready for planting the leeks** (R), allowing them to grow with their root way down, improving the blanching of the stalks (both the quality and the length), and making future hilling of these valuable crops easier.

DIY or Buy 43

Garden Operation Cycle

There is one thing for sure: if you don't know where you are going, you won't get there! Another sure thing: know the steps to get there if you want to arrive successfully. This is good advice for gardening. **Start-up** growers understand they want to grow a garden, but defining that they want a *successful* garden and *what success looks like* in terms of variety, yield, quality, and timing is what will help them **scale-up**. **Pro growers** know that most of their work falls into a *series of steps*—tasks that follow each other logically. Successful growing is very much about anticipating the needs of production for each step and meeting them with know-how and the right tools for the job. In this sense, pro growers are masters of timing, tools, and techniques. Let's first understand the importance of a step-by-step approach, and in "Part 2: Your Garden Operation Cycle," we will break these stages down into major garden tasks.

Best Management Practices

There are many standards of garden production: Organic Farming, Ecosystem Design, Biodynamics, etc. Most approaches have an underlying list of *best management practices*, tried-and-true and commonly accepted approaches to gardening. Best management practices should inform our garden operation cycles while also allowing adjustments for innovation within particular systems like Permaculture, Biodynamics, succession planting, etc.

Following the garden operation cycle explained in this book is a tried-and-true approach that makes use of well-accepted and innovative tools and techniques for best results. It is important to know how to *simply* achieve a garden task, such as seeding, in the best way possible. However, there is more than one way to accomplish any garden task—from seeding to weeding. For instance, *straight rows* remain a best management practice, a time-tested way of doing things. Straight rows lead to easy weeding, proper spacing, consistency in records, ease of laying drip, etc. At different scales, what changes is *how* (the tool and technique) not the *why!* The tools you use to accomplish any task are part of the decision-making matrix for your scale.

Tool, Technique, Timing

Every production stage has a measure of best management success. Successful gardening is all about the *right tools* for the task, the *proper techniques*, and *optimal timing*. For instance, consider using a broad hoe (tool) to hill potatoes (task) by passing on either side of the row and pulling soil into the row for 2"–3" coverage (technique) every 2 weeks (timing) until the plants begin to die back and the harvest production stage starts. The **right tool** is the one that meets any *weak links* in your production stage. For example, if you are bending over to hand seed one seed per inch in five rows, each of which is 25 feet long, it becomes clear that seeding is a weak link in your operation; a push seeder may be the right tool for you. Each tool, like a push seeder or collinear hoe, has its **proper techniques**, such as the steady foreword push of the seeder without tilting the hopper too much (too much tilt, and the seed won't get picked up in the hopper). **Optimal timing** is also crucial. If you seed your crop and it doesn't rain for a week, it would've been optimal to have used a sprinkler to help germination within a few days of seeding to keep everything on schedule.

A Step-by-Step Process

When I first started gardening, I went headlong into it. I had a sense of what to do and when to do it, but honestly if it weren't for the snow on the ground, I might have tilled my garden before reading the seed catalogs. It is now obvious to me that I needed a sense of what I wanted to grow well ahead of preparing the land. In fact, I would have been wise to *first* analyze the soil to understand what it *could* grow even before I started reading those seductive seed catalogs. Then later, I would have planned my garden beds and figured out how to seed, weed, water, etc. Gardening success is a step-by-step process. The time, money, and energy saved through taking the logical seasonal steps are calculable: you save money because you can order exact seed quantities, and you save time because you only have to prepare the right amount of row feet of garden bed for your crops. When we understand these steps, we can see how tools can help us save time and energy, and what is more, we can make sure to buy and make the right tools for each step.

The **garden operation cycle** is the entire organization of step-by-step processes for a successful year of garden design, planning, preparation, planting, harvest, and more. Within the four seasons, the garden cycle is divided into **production stages**, like doing *greenhouse starts*, and subdivided into **garden tasks**, like seeding the vegetable starts or potting up the transplants. The pro grower knows their garden operation cycle like the back of their hand … or, really, like the front of their spreadsheet. A garden plan will state every stage and show which crops are to be grown where and how they will be grown using which tools and techniques and what timing.

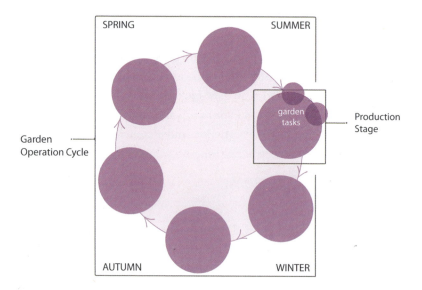

This step-by-step approach is the most powerful concept in gardening! The gardening season can make you feel like you are navigating an MC Escher labyrinth staircase; you will face many obstacles that from the outside look difficult, but, by following the rhythm of a strict garden operation cycle, each step is a successful task leading to a landing where the next production stage is achieved. Eventually, we come full circle to a new year again.

Production Stages and Garden Tasks

Production stages are periods made up of many garden tasks that work together to meet an important and often season-specific garden operational need: seeding, weeding, harvesting, etc. There is no set number of stages, though in this book I present them as 16 stages. Sometimes a production stage occurs only once, such as **primary land preparation**; other times, a stage repeats only once per year, such as **greenhouse starting**; or, routines may be ongoing, like **seeding and transplanting**. Garden **tasks** are any short-duration job, like row marking, calibrating your seeder, or seeding the rows, that must be done in an orderly fashion, whereas a production stage usually occurs within a timeframe of weeks or months.

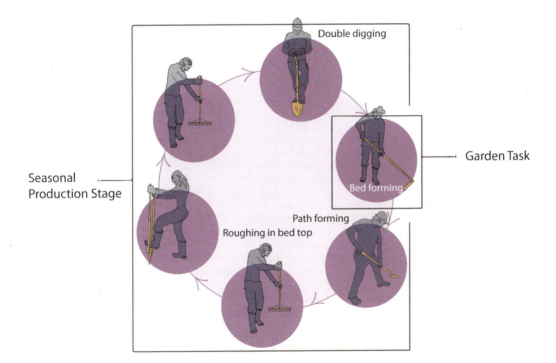

FOCUS

The Garden Operation Cycle and Production Stages

Here are the 16 production stages in the garden operation cycle. I've included some key garden tasks for each production stage and listed a primary tool needed for that stage.
Note: Sometimes a scale-up tool variation is suggested for the given stage or for different environments or enterprises.

1. Site Analysis and Sampling
   - **Soil sampling future plots** ⟶ Soil probe
   - **Aerial photos of whole site** ⟶ Drone
   - **Measuring future plots** ⟶ Measuring wheel
2. Garden Design and Crop Planning
   - **Crop planning** ⟶ Spreadsheet software
   - **Garden layout map** ⟶ Aerial images
   - **Seed order** ⟶ Seed catalogs
3. Garden Starts
   - **Seeding** ⟶ Propagation tray system
   - **Growing out starts** ⟶ Grow light and Shelves
   - **Potting up** ⟶ Various-sized Pots and Cells
4. Primary Land Preparation
   - **Breaking new ground** ⟶ Pick and Hoe
   - **Improving soil fertility with compost** ⟶ Dump cart
   - **Double-digging** ⟶ Short-handle spade
5. Plot and Permabed Forming
   - **Forming bed shoulder** ⟶ Broad hoe/Grub hoe
   - **Subsoiling** ⟶ Broadfork
   - **Roughing in the bed top** ⟶ Rake
6. Fine Seedbed Preparation
   - **Top-dressing compost** ⟶ Bucket/Wheelbarrow
   - **Tilthing** ⟶ Rake/Tilther
   - **Pre-weeding** ⟶ Tine weeder or Flame weeder
7. Seeding and Planting
   - **Row marking** ⟶ Row-marking rake/Dibbler
   - **Seeding or planting** ⟶ Seeder/Transplanter
   - **Row labeling** ⟶ Wooden labels and Markers
8. Irrigation
   - **Pumping** ⟶ Pump + Filter
   - **Field distribution** ⟶ Transport lines + Manifolds
   - **Plot irrigating** ⟶ Sprinkler lines + Drip lines
9. Garden Crop Maintenance
   - **Trellising** ⟶ Trellis stakes and lines
   - **Pests/fertility** ⟶ Backpack sprayer
   - **Mulch** ⟶ Chip mulch fork/Weed barrier
10. Crop Weeding
    - **Blind weeding** ⟶ Tine weeder/Tine harrow
    - **Between-row weeding** ⟶ Collinear hoe/Stirrup hoe
    - **In-row weeding** ⟶ Wire weeder/Finger weeder
11. Garden Harvesting
    - **Leafy greens** ⟶ Harvest knife/Greens harvester

Garden Operation Cycle

- **Carrots** ⟶ 4-tine digging fork/6-tine fork
- **Transport** ⟶ Crates/Vermont-style cart

12. Season Extension
- **Crop protection** ⟶ Row cover
- **Variety selection** ⟶ Garden journal records
- **Hothouse** ⟶ Hoops and Plastic covers

13. Post-harvest Handling and Curing
- **Leaves** ⟶ Manual salad spinner/Electric spinner
- **Roots** ⟶ Root spray table
- **Garlic curing** ⟶ Stackable trays and Fans

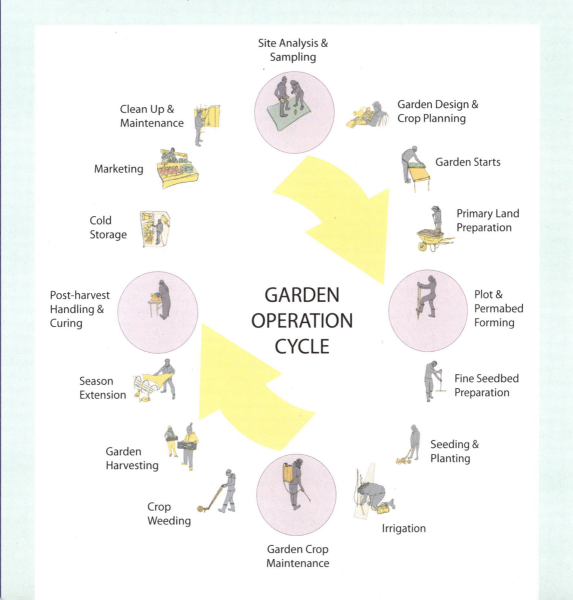

14. Cold Storage
- **Storage container** —> Rubber bin
- **Cold Temperature** —> CoolBot and Air conditioner
- **Condition records** —> Record sheet and temp/Humidity gauge

15. Marketing
- **Table** —> Folding table
- **Display** —> Display rack
- **Signage** —> Laminated signs

16. Cleanup and Maintenance
- **Space cleanup** —> Broom and Vacuum
- **Tool maintenance** —> Oils, Brushes, and Sharpeners
- **Storage** —> Tool wall holders
- **Garden cleanup** —> Cover crop broadcast seeder

Note: For most of these categories, growers at a certain scale would consider two-wheel tractor attachments or other equipment.

Tool System Design

When we see the garden as a seasonal operation cycle with many production stages and tasks to do, and we understand it is a step-by-step process, we are thinking about the whole and the parts. This is *systems design*, in which many pieces make up a functional whole. **Tool system design** is understanding how multiple tools work together to achieve your successful production stages.

Tool Guilds

To better conceptualize tools working together, I use the term *guild*. A **guild** is an assembly of three or more units (plants, enterprises, or even tools) that work together to achieve something bigger. Garden tasks often require more than one tool for completion. A **tool guild** is a selection of tools working together to complete a series of garden tasks. For instance, building Permabeds (raised garden beds) requires digging and raising the soil material into long, mounded beds. This can be accomplished with a *bed-forming tool guild:* spade + grub hoe + garden rake. Together, they achieve much more than they can alone. This is what is meant by a "guild." The **spade** can double-dig the earth and dig out the path, the **grub hoe** pulls material into a raised bed, and the **rake** can rough in the bed top and do finer finishing by pulling stones and clods into the path.

Tool Guild — D-handle Spade, Grub Hoe, Wide Rake

Tool Systems

Tool systems include all the necessary tools to achieve a production stage without compromise, having been designed to cover all bases. You may feel you instinctively know which tools form a guild that will get your jobs done, but if you sit down with pen and paper, you may be surprised. Take the time to assign tools for every garden task and form a routine of tasks for each production stage. The result is a tool *system* that can begin to function as a *whole machine*. For example, propagating garden starts requires specific tools for germination, growing out under lights, heating, watering, etc. An irrigation system needs all the components to pump, distribute, and irrigate. There are many ways to manage transplanting your vegetables into the field and lots of tool options, but the Paperpot system is its own unique system requiring specific trays, paper cells, and transplanter tools. *Systems* are different from *guilds*; they require a number of components to function, or they are assembled with great intention for the use of one tool after the other in a step-by-step process to complete a production stage or multiple stages in unison. For instance, row marking multiple straight rows on a bed top with a between-row spacing that is compatible with the width of the hoe you will use to weed, or dibbling in-row spacing at the right spacing for the finger weeder on your wheel hoe.

Pro-Tip: Yes, tools can have multiple functions across different production stages, like using a garden fork to clean up debris and subsoiling a garden bed, but as you **scale-up** you may find specializing your tools helpful, perhaps opting for a digging fork or broadfork for spring garden bed aeration, and a compost fork or more specialized garden fork for cleaning up bed tops in the fall.

Complete Tool System

Most garden tasks in your seasonal operation cycle can benefit from a tool or two to improve efficiency, ergonomics, and overall productivity. When a grower has a tool, technique, and timing for all garden tasks and the entire *operation cycle* is performed without weak links, then we could say the farmer has a **complete tool system** for their current garden operation cycle and garden/farm scale. A complete tool system for your property or farm entails all the tools needed for your whole operation cycle, ideally addressing all weak links for your given scale of operations. If you increase production, new weak links may emerge, and your tools system won't feel complete because it may be under-powered for certain production stages.

Note: Every homestead or farm will have weak links—garden tasks that are not being performed optimally—in their garden operation cycle.

Tool System Design 53

TOOL VIEW: Tool System

This collection of tools represents a practical set of **tool systems** for several of the key field production stages: Primary Land Preparation, Permabed Forming, Seeding and Planting, Crop Maintenance, Crop Weeding, and Garden Harvesting. If we added in here the rest of the tools needed for the other production stages, we would have a **complete tool system** for the farm.

A. measuring wheel; B. push seeder; C. wheel hoe with stirrup hoe blade; D. backpack sprayer; E. rubber mallet; F. painted field row markers; G. garden fork; H. socket wrench, multi-screwdriver, and tape measure; I. harvest knife, trowel, transplant ruler, and pruning shears; J. five-row dibbler; K. stirrup hoe; L. all-purpose hoe; M. rubber bins; N. harvest crates; O. tine weeding rake; P. row marker for garden rake; Q. bed preparation rake; R. broadfork; S. tool kit bag; T. step-in-fence post for plot layout; U. landscape flags; V. long-handled spade/shovel; W. seeding tool kit; X. grub hoe; Y. Clevis pick hoe

Tool Kits

When tools are chosen to work together, we call it a guild conceptually, but when many small tools are assembled in a bag or case, we call it a *kit*. **Tools kits** are collections of essential tools in a handy grab-and-go bag. Kits are great for larger systems, like irrigation or high tunnel maintenance. With your kit bag, you always have what you need when you need it and never have to make unnecessary trips back to the barn or garage. Once kits are assembled, they will continue to build themselves—any time you find you need a tool when doing a job, add it to your kit. If, for instance, a ¼" socket wrench is often needed for a job, then add one to the kit as a fixed entity. I keep a socket wrench set for general barn use and buy individual wrenches for any tool kits that need specific sizes.

Irrigation Kits

Irrigation kits are very useful because there are many fittings and tools required, and they easily fit into kits to keep you organized in your shop or out in the field. I have three different irrigation kits: an **irrigation tool kit (1)**, an **irrigation large fittings kit (2)**, and an **irrigation small fittings kit (3)**. The tool kit has a **torch** (A), **hex drivers** fit to small hose clamps (B), **socket wrenches** fit to larger pipe clamps (C), **adjustable wrenches** (D), **metal snips** (E), different-sized **drip punchers** (F), a **punch-n-cut** that can punch holes, insert emitters (¼"), and cut polyethylene tubing hose punches (G), larger-diameter hose and PVC pipe cutters and backup **cutters** (H), **pipe wrenches** of different sizes, including lightweight aluminum (I), serrated **fisherman's knife** (J), **pipe saw** (K), **multi-tool** (L), and **work gloves** (M). The fittings kit carries the items I need most often (more fittings are stored in the barn—see organization section 16. Cleanup and Maintenance. These include small **hose clamps** (N), large **clamps** (O), essential **drip tape fittings** (P) such as insert barb x drip tape, drip tape shut-off valve, couplings, end plugs, etc. There are also **extra parts** for any **sprinkler systems** (Q) I work with, like Xcel-Wobblers for larger plots and Fast-N-Fast Dan sprinklers for small plots, Shrubbler sprinkler heads for container gardens, and drip stakes for pot production. For **small fittings** (R), I often carry bits and pieces to make brass manifolds, and all the **spare**

Tool System Design   55

parts for the sprinklers, like sprinkler nozzles, male and female connectors, goof plugs, and **PTFE** tape (S).

**TOOL VIEW:
Irrigation Kit**

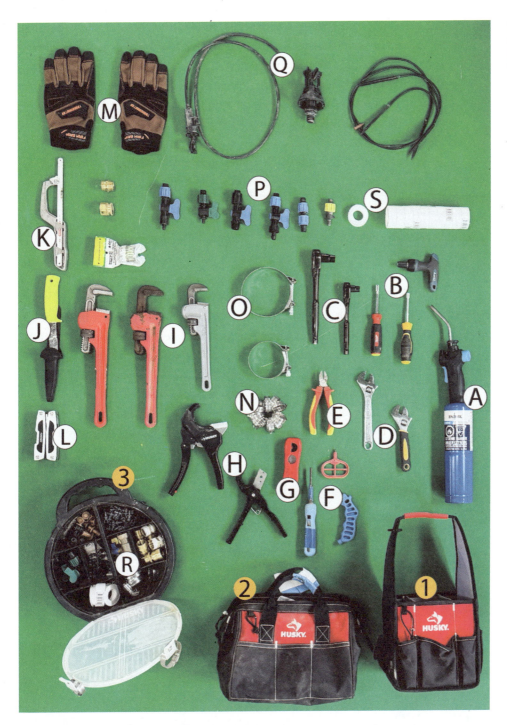

Scale-Based Decision-Making

Scale is one of the most important concepts in garden and farm planning. However, scale is an elusive concept, and worthy of better understanding.

All gardens and farms have a **scale**, which can be measured in many ways that affect our decision-making concerning tools, techniques, and timing (see decision-making matrix, below). The size or acreage of the garden, how intensively or extensively you are going to work the garden, and whether your intention is homesteading food production or commercial sale are all important. Let's consider some of these **principles of scale** and how they alter tool choices and their techniques, (though never practices).

Scale-Based Factors Affecting Tool Decision-Making

- **Enterprise Type(s):** Growers focused on garlic will invest in tools for dry crop curing. Salad growers will focus on tools for leafy crop washing.
- **Environment, Soil, and Climate** will all effect tool choice. For instance, soil type (loam, clay, sand, stony/rocky) has a huge impact on which types of hoes you can use effectively. A stony soil will benefit from more aggressive hoe types, such as a stirrup hoe with an oscillation feature that allows stones to slip between the hoe blade and handle. A very rich organic soil with few stones can easily be managed with more fine-bladed hoes, like the collinear hoe. For a light organic or sandy soil, a tilther will work well for seedbed preparation (see later section); but a tilther wouldn't be a choice with heavy clay soils because it lacks the needed power to turn the soil.
- **Actual Production Acres** is the amount of land used to grow a crop, whether it be 100 ft^2 within a 1-acre property or 1 acre within a 5-acre property. The first garden will need far fewer technical tools and techniques than the 1-acre garden to be successful.
- **Intensive Garden Management** means tighter row spacing and more turnover (succession planting) of garden beds in the growing season, whereas **extensive management** often entails wider row spacing and less bed turnover. Overall, intensive gardens yield more,

high-value crops in the same period but require more people hours invested. When it comes to tools, these two different approaches will require different choices. For weed management, the intensive grower will be very reliant on pre-weeding and blind weeding techniques using tine weeders, and the extensive grower will favor tools like wheel hoes. Also, commercial growers are more likely to invest in more fine-tuned tools that are justified with the income from high-quality, high-yielding crops than homesteaders who are growing for their own use.

- **Infrastructure** may have consequences on the space available for tool storage. Use of some **equipment types**, such as tractors, may render some tools less useful. The types of **inputs** (like fertilizers) and **outputs** (like large quantities of melons) will also affect tool choice—like how to apply the fertilizer you desire and how to move the melon loads. Finally, consider that **profitability** will always ask that we weigh and balance the costs of tools and their operation-saving and income-generation potential.

Asking yourself how each of these principles might affect your tool choices is applying **scale-based decision-making**.

Pro-Tip: It is important to remember a homesteader has savings to gain from growing like a pro that can be considered on the same level as a commercial grower. Putting aside 300 lbs of sweet carrots every winter for a family of five saves hundreds of dollars and offers much higher nutrition than store-bought!

FOCUS

Principles of Scale for Decision-Making

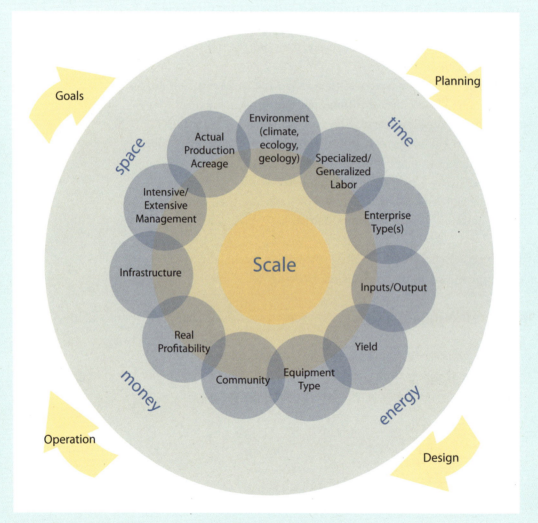

These principles of scale should be considered when making decisions about tool choices.

Scale Phases

All market gardens, homesteads, and small farms go through *scale phases*. **Scale phases** are stages of significant change in your operations. They are mostly measured in an increase or decrease in land under management, the number of times garden beds are planted with new crops, and overall, the increase in quality and quantity of produce yield. The most basic division of stages is **start-up**, **scale-up**, and **pro-up**. Each phase brings you to a **static scale**, that final measure of productivity when you are no longer expanding but holding a steady state of garden tasks within a set acreage and using a set list of tools for your scheduled tasks. ***Remember***: Your production acres or tool inventory at the start-up or pro-up phase has little to do with some *set-in-stone ideal* and more to do with your chosen *static scale*. If your final acreage is ¼ acre, your pro-up phase will look much different from that of a grower whose final acreage is 5 acres!

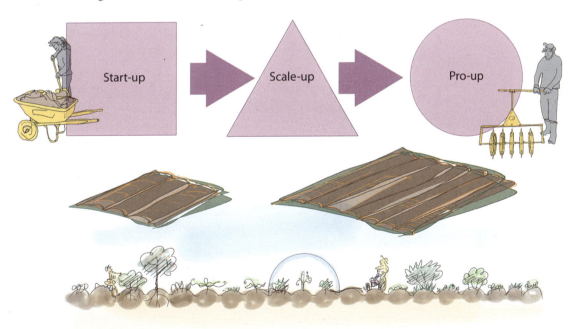

Start-Up, Scale-Up, Pro-Up

Each scale phase has a character of its own and tools that fit the job. As we scale-up through the phases, we choose tools to improve efficiency for garden jobs and those that perform unique tasks necessary at larger scales of operation. It is important to note that this concept isn't the same for all scenarios. For instance, a market gardener may decide to start-up with very different tools and even include a two-wheel tractor as part of their early acquisition of equipment. A home grower probably wouldn't make those same choices. The types of *tasks*, though, that early equipment purchases are focused on stay similar across different scales of operations—from home to professional growing.

So, what does it look like when you have fully reached your **static scale** as a pro grower? Usually, you will have scaled-up your systems for each garden task in your operational cycle 2–3 times (maybe more). This doesn't mean you went through 2–3 complete overhauls with all new tools! It means you addressed the **weak links** in your tools, techniques, and timing to get where you want to be—your **planned static scale**.

- **Start-up:** When you first start-up your garden in years 1 to 3, mostly acquiring tools that help with garden plot and bed layout, forming and finishing, as well as simple management of essential basic garden tasks, like seeding and weeding.
- **Scale-up:** Often, in years 4 and 5, we streamline our garden projects and acquire tools that maximize efficiency and add key stages to our garden operation, such as pre-weeding tools (like tine weeders) and improved seeding capacity. Irrigation is a priority in this phase.
- **Pro-up:** By year 6, you will mostly have turned your garden to the pro-up phase. Tool acquisition will mainly be focused on fine-tuning operations and bringing in tools that maximize efficiency at your final acre scale (a multi-row seeder, perhaps). New tools and techniques are particularly valuable for harvest and post-harvest. A pro grower may find that multiples of the same tool help them get time-sensitive tasks done at the right time.

Scaling-Up Tool Systems

As you scale-up to your static scale, you will need to scale-up your tool systems. Indeed, tools are one of the key ways we scale-up from smaller home gardens to larger homestead productions or, as market growers, from start-up to pro-up commercial production. Yes, we may increase our actual production acres, but ultimately, tools are what save us time and energy and improve quality—saving and making money. Tools are a key way of scaling-up, but we have to do it carefully because the right tool can help us, and the wrong tool can push our production down a path we may not wish to go on. For instance, if you buy a seeder that seeds tight row spacing, but you want to grow more extensively, you will find it becomes obsolete and a poor investment.

Scaling-Up Your Tools

For every task there is a tool, and, depending on your scale of gardening, you may choose to scale-up or even pro-up the tool being used. This may entail going from a single-row dibbler to a 3-row dibbler. The technique is often the same, but the tool used for larger gardens or more intensive gardens will change. In a large garden, say 300 ft^2, you will need to accomplish tasks more quickly, and for gardens that are more intensive, with closer row spacing or more successions per season, you will want tools that are more efficient and more able to be fine-tuned. This doesn't mean the previous

Scaling-Up Your Tools

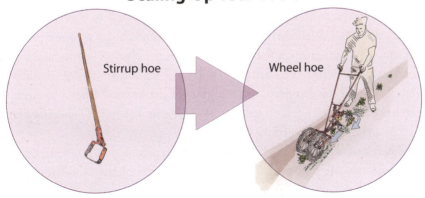

For a start-up grower, a stirrup hoe is a great tool for between-row weeding. As you scale-up, a wheel hoe can increase the pace of weeding in the paths and bed tops for wider row crops, and added attachments (like finger weeders) can help with in-row weeding too.

tool becomes obsolete, but rather the context when it is used is expanded to include tool versions that are more useful in some of the applications. Consider, for instance, the stirrup hoe used to weed the bed top and paths. Scaling-up, a grower may continue to weed the bed tops with the stirrup hoe but work the paths with the faster and stronger wheel hoe.

Scaling-Up Your Production Stages

Tool systems are best focused on the *stages of production* in your seasonal operation cycle. The weak link tends to be the first challenge growers face as they begin their garden careers or enter a new phase. **Start-up** growers need to prepare land better; **scaling-up**, you'll want to improve seeding, weeding, and watering; and **pro-up** growers will focus on harvest, post-harvest handling, and cold storage. This is the typical trajectory for many growers—probably including you!

Weak Links

Production stages form the backbone of a successful season when each is achieved in order. This makes them a practical framework for tool decision-making as you start-up, scale-up, and pro-up your garden or farm. It is easier to make decisions based on a production stage rather than the smaller tasks or the whole operation cycle.

There will inevitably be weak links in your garden operation cycle—tasks that are not being performed optimally. As we scale-up, we will push the limits of certain tools for tasks until one task or another becomes the weak link in the production stage. For instance, I used to dibble my garlic beds with a single-row dibbler that pressed six holes at once in only one row. As I scaled-up, I found I needed to get this job done faster and more precisely across the entire bed top, so I shifted to a rolling dibbler. However, I still harvest garlic with digging forks because I have consistent volunteer labor every fall to help with the garlic harvest, so there was no need to change this part of my tool system.

When looking for weak links, look at which production *stages* are slowed down or have quality issues. You may miss the big picture and forget how the stages affect each other if you start small—nit-picking single tasks for issues in efficiency. For example, you may feel you have poor carrot yields. Don't assume this is a *harvest* issue because harvest-damaged carrots are reducing yield, and don't be certain that *large weeds* are reducing yield because summertime weeding isn't adequate. Perhaps it is actually that the germinating carrots are outcompeted by early weeds and the **weak link** is solved with pre-weeding (preparing the seedbed and allowing weeds to germinate and killing them easily before even seeding the carrot crop). The **tool solution** may be a tine weeder or tine harrow for better stale seedbed preparation.

Scale-Up Production Stages Relative to Each Other

Many production stages need to be scaled-up in unison to avoid creating *new* weak links. For instance, if you jump in and scale-up your salad harvest and post-harvest production equipment with a greens harvester, salad spinner, and a larger wash basin with aerators, you may be tempted to grow more salad to meet this infrastructure capacity. If you then find your overall capacity to produce quality salad greens in the field is hampered because you can't pre-weed well enough, you may have scaled-up one stage too fast relative to another. You should always scale-up a tool system relative to other stages. If you find you can easily manage weeding, then you may be ready to scale-up other stages without putting pressure on any one task. Why is this important? Scaling-up can become a slippery slope of just jumping up ladder rungs without thinking and then adding tools to solve problems.

Pro-Tip: Don't buy tools to solve problems! Make strategic purchases and try DIY constructions to solve the weak links that are obstacles to the efficient operation cycle that moves you toward your intended static scale.

As I scaled-up my market garden, hand tools played a critical role in field management. The wheel hoe was key because it can move quickly through my larger fields (and it's a great upper body workout!).

Scaling-Up Your Avatar

Every grower follows a *type of production*, an enterprise—like market gardening, orcharding, or beekeeping. Sometimes growers run multiple enterprises or **Guild Enterprise Production**, in which several enterprises work together, sharing space, resources, and tools. As individuals running our own enterprises on different pieces of land, our guild enterprises will all be unique. Let's visualize growers as unique *avatars* running their enterprises. Using this *avatar concept*, see if you can find yourself within the examples provided over the next pages.

Find Your Pro Grower Avatar

Let's define some of the more popular avatars.

Home gardeners grow food for their own use in their own yards. Garden sizes are usually small, and usually the only enterprise is vegetables. **Homesteaders** try to grow a larger percentage of their foods from home and engage in many DIY projects on their property, adding more enterprises, like edible landscaping. **Back-to-the-Landers** have returned to the land *as a way of life* and grow, raise, and make many items for themselves through several enterprises. They often engage in some commercial sales of products on the side and barter extensively. **Small-scale farmers** grow and raise food for sale, usually with a focus on a selection of complementary products for retail or wholesale. **Market growers** have a known focus on market garden vegetables for direct sale at farmer's markets and grow a lot of varieties of annual crops. **Urban growers** maximize small spaces for very high yields with intensive practices.

Your Production Practices

You can start-up, scale-up, and pro-up as any avatar, and enterprises can be added or removed. But I recommend a focus on three core enterprises for a successful land management system. Remember, however you grow, using Organic, Permaculture, Ecosystem Design, and other best management practices can be followed by any avatar's enterprises, like **Organic Farming** or being a **Permaculture grower** (see *The Permaculture Market Garden* book).

Scaling-Up Your Avatar   65

Edible Ecosystem designers use specific methods for growing highly diversified and organized annual and perennial garden systems for homes and homesteads using Permaculture methods, market grower techniques, and more (see my *The Edible Ecosystem Solution* book for more on this).

Pro Avatars

Understanding your avatar can help you scale-up and make the best choices for tools, techniques, and land management. If your avatar is a home gardener, you will make different choices than a market grower would. The pro-version of your home garden avatar wouldn't consider the same tools as the pro-version of a market grower (which is a commercial enterprise). A start-up home gardener will be getting their bearings; scaling-up, they will be setting their course, and as a pro home gardener, they will be in their stride. In all **scale phases**, however, they will be a home gardener with the correct tools for *this* avatar. However, as growers scale-up, they sometimes shift their avatar—becoming a market grower or a homesteader, even though they started as a home gardener.

Avatars

Top row left to right: *Suburban Permaculture Homesteader, Back-to-the-Lander and Small-scale Farmer*

Bottom row left to right: *Market Grower, Homesteader Start-Up to Scale-Up*

66 *The Garden Tool Handboook*

FARM FEATURE
Building Urban Beds at Northern Liberties Community Garden

There are many ways to build a garden—using many different tools. Often, primary earthworks involve tractors (even for plots that eventually will be worked with only hand tools) because the power of a two-wheel tractor to open new ground is so great. However, successful turnover of new land can be achieved with enough labor, a tape measure, a spade, a broad hoe (or grub hoe), a shovel, and a rake. These represent an essential tool guild for start-up growers—and all that was needed for the primary land preparation and bed-forming stages of this urban community garden plot. *Sometimes more hands and simple tools are all that is needed to achieve the desired result.*

Building new Permabeds as part of rejuvenating an urban plot in a Philadelphia community garden.
(1) assessment and design. (2) double-digging. (3) forming new beds. (4) roughing in the bed top and finishing the bed top.
(5) producing an urban Permaplot that maximizes space for soil and minimizes wasted edges by combining the stones and bricks into one thermal wall that can receive pepper pots.

1

2

3

4

5

FOCUS
Essential Tools Stick Around

Over the years, every grower will acquire a collection of essential tools in their shed. Our goal is to make good tool purchase decisions to ensure we have the best tools possible for our garden operation cycles. We do this by selecting tools specifically for each production stage and the garden tasks within each stage.

Over the years, some tools will get used more than others and *stick around* in our shed in a ready-to-use state, while others may be used less frequently or eventually not at all. Almost always, the reasons for a tool's lack of use are preventable. Reasons include: (1) tool is not suited to the grower's scale and is either over- or under-kill for the garden task, (2) tool is not suited to the garden task it was purchased for, (3) the grower isn't clear about proper use of tool, (4) the grower scaled-up quickly and the tool is obsolete, (5) tool eventually wears out and isn't repaired. Occasionally a tool may be of poor design for its intended purpose (6), and so not used, or (7) the tool wasn't well-made and breaks or wears prematurely and is never repaired.

A tool from a great tool company, and especially tools designed for professional growers, are less likely to fall into category 6 or 7. If they do fail in some way, customer service will often remedy this and use the information you give them in future re-designs of the tool to improve it.

For the other categories, consider the following solutions discussed elsewhere in this book: (1) understand different production scales and tool options, (2) learn garden task types and tool options, (3) learn proper tool techniques, (4) set a future static scale using scale-based decision-making, and (5) learn practical storage and maintenance for tools.

Part 2:
Your Garden Operation Cycle

Now, let's look at your garden operation cycle by considering an idealized series of seasonal production stages and garden tasks viewed individually in relationship to the tool options for growers at different scales. These 16 stages (which you have already seen in the Part 1 Focus box: "The Garden Operation Cycle and Production Stages") are not set in stone, but they are a practical way of looking at the whole operation cycle in discrete sections from the point of view of tools, techniques, and timing.

1. Site Analysis and Sampling

The first stage is the analysis of your site, your garden, property, and field and the sampling of your soil, weeds, water, etc. Every garden sits within its environment, and we must understand the slope, drainage, soil texture, sun exposure, ecology of the plant community, and climate to grow better food.

Environmental Site Assessment

Site assessment comes from observing the natural environment of your site. Sampling can help with this assessment, and the analysis of all collected data should guide the placement of your new garden. List numbers correspond to the collage.

1. **Determine your elevations** to help site a garden in a flat or gently sloping terrain or decide if terracing is necessary for land with steeper terrain. Use **real topographic maps** to view your local terrain, or toggle on the topography feature with popular **online map tools** and zoom into your county and property on **interactive maps**. In Ontario, the Ministry of Natural Resources and Forestry has a site that allows you to make a topographic map. You search your address and toggle on imagery layers to see aerial views of the land and vegetation and then topographic

1. Site Analysis and Sampling   69

layers to see the contour lines (lines of equal elevation). Of course, you can (and should!) walk your land and take notes to see how this plays out in 3D as the wolf trots. Your goal is to find areas that are suitable for gardens, topographically speaking. Gentler land turns into garden more easily; too low is too wet to garden, steeper land requires contour planting. If the land is very rocky and steep, it is best to focus on perennials like orchards, medicinal meadows, grazing lands, etc. You *can*

This simple topographic map tells you so much. As you can see, the elevation is between 135 meters to 151 meters above sea level. At the bottom of the map, the distance between the contour lines (135 m to 140 m) is about 150 to 200 meters as the gardener walks. This is one of the gentler areas. On the other side of the steeper ridgelines (up to 151 m), there is wetland (note the wetland symbols). By actually walking the land (and viewing the aerial imagery), we learn the following: the wetland is very wet and full of reed canary grass; the hillside is full of large hardwood trees and large rocks; and the soil is a stony loam. Soil sampling shows the bottom field is a clay loam with fewer stones closer to the road, and it's a little wet in spring as you move 300 meters from the buildings (bottom left of map). Once soil samples, field walkabouts, aerial imagery, and topographic features are studied and considered together, you might come up with the following plan of action: build a shed on the highland, where there is good support and drainage; plant perennial food forests on the ridgeline, between 150 and 140 meters; and plant gardens and crop fields in the gentle land between 140 and 135 meters. Wetlands marked on the map are to be dug out to become irrigation and duck ponds, and wet areas in other fields will be dug out as linear bioswales between garden plots.

terrace this land to create gardens, but a great amount of labor is required, and you need a proper design for elevations and terrace widths that accounts for tools/equipment scale.

2. **Identify your current ground cover** by using a spade to dig up the current sod or weeds to determine if they are annual or perennials. You can identify them with weed books or plant ID apps. You can also look at the roots of grasses to see if they are fibrous (indicating an annual) or thick and succulent (a sign of grasses like *Phalaris arundinacea* that resprout aggressively from their roots). Perennials will persist even after cultivating the ground; they need to be fully removed or killed back completely. Aerial imagery will also show distinctions in plant species. In the photo, the large swathes of reed canary can be seen when correlated with data from the ground. For instance, with an aerial image in hand, you can see area of different shades of green in an old hay field. If you inspect those areas on foot, you can positively ID the reed canary grass in different area within the perimeter of the aerial image's suspected areas. Once this is confirmed in enough places on land, you can make assumptions about the overall distribution of the species across a field and better plan for its removal for new gardens and food forests with a plow/cultivation/cover-crop work plan. Once samples are taken in the field, the extent of the weeds can be outlined in the aerial images.

Pro-Tip: *Drone imagery will be even more accurate in real-time, but doing a field walkabout is essential to confirm imagery results.*

3. **Discover your climate, hardiness zones, and elevations** by looking up your ecoregional climate and understanding its temperature and precipitation trends. Also, research the most current hardiness zone map for your area; If you can find updated maps that have taken climate change trends into consideration, all the better.

4. **Observe the sun** to determine which areas get full sun, partial shade, or full shade. It is critical to understand how much sun a garden site receives. Sun-loving plants simply won't thrive in shady sites! Use a **sun calculator** tool to determine the amount of sunlight received in a specific location over a 12-hour period on any given day. Print out a property map or a Google Earth screenshot and visit your potential garden site in the morning, mid-day, and evening. Take notes and color in the areas that are full sun (yellow) and shady (purple). Repeating this process throughout the growing season is necessary to make a proper assessment, but if a site is seemingly in full sun in the spring, that is a

1. Site Analysis and Sampling 71

Site Assessment Tools

Your garden needs:

Soil pH Adjustment
Soil pH is low. Apply 5 pounds of lime per 100 square feet and be sure to incorporate the lime into the upper 6" of soil. See information sheet.

Nitrogen (N)
You will need a total of about 3.5 ounces of nitrogen for every 100 square feet of garden for the entire year. Soil organic matter content will help lower this amount/rate by providing nitrogen. It's best however, to apply this amount throughout the growing season. See information sheet for calculating rate per application and suggested timing.

Phosphorus (P)
Soil P level is relatively high. Broadcast about 1.5 ounces of phosphate (P2O5) per 100 square feet prior to planting and work into the upper 2-4" of soil. Apply a starter solution high in P when setting transplants. See information sheet.

Potassium (K)
Soil K level is relatively high. Broadcast about 1.0 ounce of potash (K2O) per 100 square feet prior to planting and incorporate into the upper 2-4" of soil. Apply another half an ounce at planting. See information sheet.

probably good site for growing. If, in the hotter months, some shade is cast by trees, you can take advantage of this and grow tender greens. Use drone imagery to get more real-time visuals of your site. Aerial imagery from a drone will clearly show if your site is shaded by trees or if it is an open space between trees; sometimes you can determine the shade effect from adjacent trees. But to truly understand if a site has full sun most of the year, you need to be on site, observing it in spring, summer, and fall. Pro growers will study detailed **hardiness maps**.

5. **Determine soil types** by referencing soil maps for your region and soil surveys. Confirm with soil samples.
6. **Sample your soil** to determine the texture, fertility, pH, and organic matter content of your soil. You can test pH with **pH testers**, and other home soil test kits. At **start-up**, sending soil samples to a lab is advisable so you have baseline data to work with. **Scaling-up**, growers will not only rely on initial soil sample data, they will do sampling every 2–3 years in the first decade of production.

It is important to understand your soil composition to determine if a site is a good option for a new garden. **Start-up** growers will use a shovel for sampling; a **pro grower** may use a **soil probe** to take core samples; sometimes this requires a rubber mallet to hammer the probe in, though some models twist into the ground. A core sample will show the horizons of the soil, the thickness of the organic matter in the A horizon, and the composition of the other horizons. To do sampling with a probe, take three or four cores and put them into a 5-gallon pail and mix them up; remove any twigs, rocks, or grass, then measure out 1 cup of the soil, and put this into a Ziploc bag labeled with the name of the plot ("Zach's Backyard Site #1," or "ZBS#1") using a Sharpie marker. This sample can be shipped to a soil lab along with a lab form that specifies which analyses you want (macronutrients, micro, pH, soil organic matter, etc.). The lab will send you the results with recommendations for any crops you specify.

You will also need to determine your soil moisture regime—wet, moist, or dry—by simply putting a **moisture probe** into the soil and reading it. You can also dig a hole in the soil and see if it fills with water after being left for 24 hours. Wherever the water fills to will be your water table at the time of the year you dig the hole. In your garden, you want your water table below the depth of working the soil (greater than 18", ideally). If your soil is dry, you can help it hold more moisture by incorporating clay and organic matter. If your soil is wet, you can improve it with sand and organic matter to improve drainage. Building raised beds will always help your garden drain better. If your soil is sandy, it is likely drier; and clay soils hold more moisture. Use a **rain gauge** to get a better sense of your seasonal precipitation and compare this to records (searchable online) of the average monthly precipitation in your region.

2. Garden Design and Crop Planning

Garden design is the whole decision-making process of placing your garden into the landscape, choosing the type and quantity of beds, their architecture, and how the garden will be managed over the long term. **Crop planning** is the process of selecting which crops you want to grow, when you will grow them, where they will be planted, and how they will be maintained each season.

Design Your Garden and Plan for Crop Success

Garden design and crop planning is a critical production stage. This stage goes hand-in-hand with site analysis and sampling; you cannot design and plan without understanding your land. As such, you will see an overlap in the discussion below. But, unlike site analysis, garden design evolves—especially over the first three years or so—and even then, it is still revisited annually. Crop planning occurs every year, allowing you to make adjustments, but the bulk of your planning will stabilize once you pro-up your production and follow the rhythm of the crops you grow and know how you want to grow them.

Turning Analysis into Design

As stated above, **start-up** growers will need to do data collection using tools like **soil probes** (A), **rain gauges** (B), **pH testers** (C), and maybe even **aerial drone imagery** (D) combined with other environmental research from hardiness and soil maps. Data affects design. For instance, if your soil is wet and heavy, building raised beds is an advisable garden design.

If you find that you have a great loam field in the right place, you'll want to flag it and measure it using a **measuring wheel and tape** (E) to reproduce the plot widths and lengths sketched out in your **design notebooks** (F); the ideal sizes will be based on your research from **books** (G). For instance, before you flag your plot, consider your Permabed architecture: if you are making 48" wide beds, then how many beds will fit? **Scale-up** your garden

2. Garden Design and Crop Planning   75

design by using digital software like spreadsheets for making garden maps on your **computer** (H) and **pro-up** by dropping your drone images into your **tablet** (I) for direct design overtop the images to lay out plots and beds. Be sure you are using real imagery that is current. For instance, you need to be able to confirm that area with reed canary grass (a very aggressive weed) so you can plan to cover that area in a poly tarp for two years while starting your garden in the less weedy soil to get a head start.

Crop Planning from Seasonal Feedback

Crop planning builds on good site design, with your annual experience serving as feedback into the design process. **Design notebooks** (F) are used to organize to-do lists, ideas, and sketches, alongside a **garden journal** (J), which should be the repository of in-season garden notes, like planting successes and obstacles. For instance, observation notes taken once per week about each crop will help you fine-tune the timing of your response the next year when pests like cucumber beetles, flea beetles, and cabbage loopers emerge. These notes will also help you remember which varieties seem to be less affected by a particular pest, which strategies you employed (row cover, crop rotation, pesticide, etc.), and which have proved most effective from year to year. When crop planning in the winter, refer to your design notebook and garden journal alongside **seed catalogs** (K) for information on variety days-to-maturity, and combine that with other **crop-specific research** (L) which can be compiled in binders from many sources: conference notes, online research, printed, books, etc. I also reference photos saved in my **phone** (M) app folders for relevant crop records (like pest images and crop health images). Once a week, I do entire field surveys using just photos and voice memo recordings (for this, I bring a **backup battery** [N]). I very much enjoy engaging with the farm in this way.

I am always scratching on **graph paper pads** (O) before putting my refined plans into my computer, iPad, and/or design notebooks. I use this same paper pad on a **clipboard** (P) in the field for casual notes and observations.

Seed Orders and Records

Seed orders rely heavily on information from seed catalogs, notes from your garden journal, and data from your actual **seeding record notebook** (Q) that has a record of all crops seeded—with dates, bed locations, how many bed feet, how many rows, the seeder and seed plate used, and other notes. This data may change your crop variety choices (for example, maybe you want to improve germination rates with a new variety). A spreadsheet with all the crop varieties you grow along one column will have all the necessary data for calculating a new seed order in the columns to the right:

the number of seeds planted per foot, the rows per bed, the number of beds you want to grow, etc. When you review your seeding records (that show which crop was seeded where and which seeder was used with which seed plate at which calibration) and compare it to your **garden journal** (which will have notes on yield from those beds), you may find that the seeding was too dense, too much thinning was needed, and the yield was also lower. So, you can revisit your seed calibration. Your **seeder calibration record book** (R) shows all your chosen seed plates and settings for specific crops. If the seeder needs changes to settings to improve crop performance, or if new calibration is needed for a new variety chosen to replace those performing inadequately, then you do a calibration session (see "Focus: Calibrating your Jang Seeder," below) and record the new settings. These records can all be combined as spreadsheet tabs on your computer, or in one spiral binder, or as separate notebooks lined up on your shelf—whatever works for you!

Crop planning also relies on financial records. Make an effort to store them in an organized way on your computer or in a **file folder** (S). You'll need good records of your income, expenses, and profits. You especially need to look at your market sales sheets and see how certain varieties performed at the farmer's market, and you'll want to consider CSA customer feedback.

Schedule Your Crop Plan

Imagine yourself at this point: your garden plots are drawn out in software, graph paper, or aerial imagery layers; your seed orders have been made with the aid of a spreadsheet or other software; your garden operation cycle is bullet-pointed in a Word document and sketched out in your design notebook; and your key garden tasks are organized visually into your crop map in a spreadsheet or some other medium. This visual organization includes notes of when to seed a crop and how many bed feet of crop you want to grow typed into the spreadsheet cell where you have mapped the garden beds as rows of spreadsheet cells. I give each cell a map scale of 4' x 5', so ten cells in a row make a 50'-long Permabed in reality. I will fill these cells in a light green and then the next bed fills them in a darker green and alternate this for the entire plot to easily differentiate between different beds in my spreadsheet map. My notes are typed into some of the cells: "Direct Seed Carrots (May 24/3 rows)", or "Turn over bed for buckwheat cover crop (June 15)", etc.

Direct Seed Carrots @ 3 rows (May-24)
Transpant Eggplant @ 2 rows and 18" spacing (May-24)
Direct Seed Carrots @ 3 rows (May-24)
Transpant Pepper @ 2 rows and 16" spacing (May-24)
Early Radish seeded (April 1)
Transpant Cucumber @ 1 row and 30" spacing (May-24)
Early Arugula seeded (April 1)
Transpant Zuchinni @ 1 row and 30" spacing (May-24)
Early lettuce seeded (April 1)

Note: If you are growing on a much larger scale (say, 3 acres or more) and your beds are much longer (say, 100 feet or more), you may decide to view the spreadsheet cell as having a map scale of 4' x 25'.

Now, is the time to put your key garden tasks onto a **digital and/or paper calendar** (T) so that you do everything you said you would when you need to! I have a wall calendar but also keep a **larger calendar notebook** (U) for staff to use, and I have my own **personal calendar notebook** (V) to keep me on track . Pro-up growers will probably use digital calendars, but they often print out relevant material for use in the field and staff facilities. When it comes to harvest days, I always write out everything on a **whiteboard** (W) (much bigger than the one shown here) that has permanent names written in Sharpie in a grid for all the crops on the x-axis and names of the sale venues on the y-axis (market, CSA, restaurant customers, etc.). Then I just add the intended harvest quantities of each crop beside each venue (using lbs, bunches, pint boxes, etc.). A **corkboard** (W) is useful for posting special reminders and opportunities for staff.

Connect to Your Ancestry

Through all this, I rely on my grandfather's and father's **design box** (X) to hold many design essentials like stencils, pens, rulers, etc. When I combine this with a **heritage shirt** (Y) and some good music and **warm tea** (Z), I find my design and crop planning sessions are very enjoyable; I feel connected to ancestry. This connection to a timescale that is multi-generational reminds me to take the time to take notes every growing season and reflect on them in the downtime in winter to make good longer-term decisions for the land and future generations.

Put Your Calendar to Use for Scheduling

All gardens must follow an **annual garden schedule**, organizing all the production stages into months and all the chronological garden tasks into step-by-step actions. This schedule is best written for each month into a calendar and followed diligently. Indeed, nothing is more valuable to a

2. Garden Design and Crop Planning

grower than a strict schedule. Even a **start-up** grower's less experienced plan is invaluable; improvements can be made in years to come. I use Excel spreadsheets for this—with templates I've developed over the years—but there are numerous pre-made calendar software programs on the market.

Spreadsheets for Crop Planning

Spreadsheets can be powerful tools for growers to create crop plans, seed orders, task schedules, and more. There are specialized apps available to help with garden planning, but readily available spreadsheet software is straightforward to use and can achieve much with minimal cost. It's all about functions.

Dan Brisebois is a market gardener with Tourne-sol Cooperative Farm in Quebec. He is a specialist in crop planning and seed saving. Some of Dan's favorite spreadsheet "functions" include *SUMIFS* that allow the user to sum the cells *if* they pertain to a certain piece of data. For example, you can have the program sum the row feet *if* they are carrots, but not *if* they are arugula. You can use XLOOKUP to have the formula reference the data on a master spreadsheet that contains, for instance, all the seeding densities for crops. The power of spreadsheets to improve farm profitability and efficiency can be testified to by the fact that Tourne-Sol has been a successful enterprise for 20 years. Cooperative farming is tricky. Managing the diversity of vegetables for CSAs is difficult, to say the least; it involves a lot of logistics. Spreadsheet crop planning made it possible for this farm to coordinate crops and improve profits annually.

As a market grower, I took scrupulous records at farmer's markets and input these into my spreadsheet crop plan, which allowed me to take my **actual market sales** from a 30-farmer's-market season and average them over the years and turn that into **harvest goals** (pounds, bunches, heads) for each crop. I could then schedule those goals as crop units needed for certain harvest dates (the Thursdays and Fridays before market weekends). Then I worked backward to determine the number of bed feet for each crop I wanted to be grown and ready on those harvest dates. I could then order seed based on **seeding rates** (from a master reference sheet) for the predicted **row feet**. In this way, my original market records helped me meet my desired yield goals for future market sales.

Crop Planning in Winter

Winter is about reviewing seasonal notes and ordering seeds, supplies, and equipment before the busy season starts.

80   The Garden Tool Handboook

The first level of using spreadsheets is learning to speak the spreadsheet language. For instance, you only put one value in each cell. One cell says *carrots*, the next says *10*, and then the next says *lbs*. Don't write: "10 lbs of carrots" in a cell! Once you understand how to divide information into cells using columns and rows, you can then move on to the next level: learning different functions to make your sheets work for you. According to Dan, the third level is mastering PivotTable, which allows you to pull information out of your sheets for powerful summaries.

If you want to go deeper into spreadsheets to master seed orders, nursery planting schedules, field planting calendars, and more, check out FarmerSpreadsheetAcademy.com.

FOCUS
Digital or Paper Design and Records

Aerial property maps are great for organizing new garden plots with a **digital drawing program** (A), and using **printed aerial images** (B) can be great for seasonal environmental records for long-term projects, like improving drainage with bioswales or planning perennial edible hedges. Creating harvest plans helps organize harvest days, but **actual harvest records sheets** (C) can (and should!) be updated in the wash area to reflect the real numbers achieved. At the market, **actual sales records** should be kept because they are critical for planning your future crops. All this data is compiled into spreadsheets to help you plan for the next year's growing season. Spreadsheet software is easy to adapt for crop planning, seed and supply orders, schedules, recordkeeping, and **garden layout maps** (D).

3. Garden Starts

Many crops can be direct-seeded into the garden when the soil is ready and weather permits. But to get a jump on the growing season, get ahead of the weeds, and even give some long-season, heat-loving crops (like tomatoes) enough time to grow in colder climates, then garden starts are the way to go.

Germinate, Pot Up, Grow Out, and Harden Off

Garden starts are plants that are germinated under the right conditions in climate-managed space. The small seedlings (aka germinates) are often started in larger, open trays to save seed and then potted up into cell packs, trays, or pots. Next, they are grown out until the desired size is achieved, at which point they are hardened off by taking them to a sheltered outdoor space or moving them in and out of the greenhouse to let them toughen up to more natural conditions of rain, wind, and sun exposure and cooler nights. Now they are ready to transplant into the field.

Germinating

Start-up home gardeners will often use their window ledges and sunrooms for plant start containers. Wire shelving units with fans, grow lights, and heat mats are common set-ups for market growers and pro home gardeners. **Scaling-up**, consider seeding into open trays by broadcasting the seed evenly and then transplanting the successful germinates at the "thread stage" into cell trays. This uses fewer valuable seeds and eliminates thinning work. It also ensures that the trays you will be taking the time and effort to water are 100% full of healthy germinates. Humidity domes can be used to maintain ideal moisture conditions for germination. **Pro growers** may consider a germination chamber built as an insulated space with temperature and humidity control.

FOCUS
Why Start Plants?

Some crops have a longer *days-to-maturity* (DTM) than is suitable for a garden's climate. Starting them indoors helps give them a boost in a short summer. Starts can also give you a jump on the season because you transplant a crop's **first succession** as seedlings rather than direct-seeding it, often resulting in a first harvest ~1.5 weeks earlier. (Market growers plant many successions, or waves, of each crop for continuous supply.) Vegetable starts, in general, facilitate a transplant-oriented approach to growing that uses fewer seeds, avoids the work of **thinning**, creates **precision planting** at exact in-row and between-row spacing (which improves weed management), and is space efficient. Some growers love transplanting any and all crops that can be started indoors. Others will stick with direct-seeding for as many crops as possible because they like the reduced workload (if they manage their weeds correctly) that seeding provides. Some crops in cold climates will, by necessity, need to be transplanted to make the most of a short season or to improve the number of successions you can achieve in a growing season (if you want to turn over a garden bed multiple times and get 2, 3, or even 4 crops in one year). The common practice for small urban growers is to do mostly transplants. Farmers working larger, more extensive vegetable beds usually do more seeding. Most growers do a mix of both. **Start-up** growers may want to outsource their transplants by buying quality starts from reputable local greenhouses.

Potting Up

Seeds are often germinated in open trays, especially long-season crops like tomatoes and peppers, and then potted up. Sometimes, shorter-season crops, like kales and lettuces, are seeded directly into cell trays. Crops like lettuce can have one or two seeds per cell seeded (not all will germinate; most varieties have between a 75% and 95% germination rate) and thinned to one plant per cell. **Pro growers** will usually seed their first crops of all garden starts into open trays and pot them up; later in the season, they seed crops like greens directly into cell trays. For expensive seed, potting up is recommended.

Growing Out

Transplants are moved out of the germination shelves and spaces and into larger areas. For a **start-up** home grower, this may be the same space, but many growers use a greenhouse to grow out the plants until they are the right size for field transplanting. The use of **bagel trays** to keep cell trays organized in greenhouses is helpful, as are potting tables. Sometimes plants are kept on the ground to reduce evaporation from air movement and/or if the ground is warmer.

Hardening Off

Home gardeners can harden off their starts outside and bring them in when it gets cold. Market growers will often harden off outdoors on pallet stacks and potting tables and use row cover for protection when needed. Some specialized potting tables have *Quick Hoops* to use with row covers and shade cloth for growing out and hardening off.

Task Flow for Homestead Garden Starting

Any grower doing starts will need space and tools set up for the numerous tasks involved. Each task has its own dedicated space (see page 85): **seeding and potting** (1), **germination and growing out** (2), **clean storage space** (3), **soil and nutrient storage** (4), and **hardening-off** (not pictured). Start-up with a **repurposed table** for a potting table (A) or scale-up by building your own out of cedar—at your best working height and with the right dimensions for your trays. This space is used to hold your **potting mix bin** (B) and your active workspace with seedling trays, like this **plug tray** (C). A catchment tray, like this repurposed **shoe tray** (D), can help keep things clean when **scooping** soil (E) to fill the tray cells. You'll need some type of **shelving** (F) to keep your most-used items (3) within reach for efficient workflow. You'll want to be able to easily reach **potting mix** (G), **mycorrhizal inoculants** (H), **fertilizers** (I), **soil probiotics** (J), **worm castings** (K), etc. Other tools to keep handy could include **sprayers** (L), **dibbler boards** (M) like this one from Two Bad Cats (these also double as a plant tray popper to help "pop" the plugs out when ready to plant), and

soil blockers (N) for this potless transplant method. The **red metal trays** (O) shown here come out of standard metal toolboxes and work nicely to keep small seeding essentials tidy. Sturdy cell packs can further organize items within the trays. Here, you can store small items: handheld seeders, in-use seed packets, widgers, and dibblers, or just teaspoons (all of which can be used to dibble holes and to pull little plants out of cells for potting up), pot label sticks, writing pens and markers, and your seed record book. Nearby, you can keep often-used **supplies** (P) like watering cans and transplant pots—peat pots, Jiffy pots, Paperpots, larger poly pots, etc.

Start-up growers can germinate on a tabletop or a shelf in a sunroom, with additional lighting if needed. **Scaling-up** growers will use wire and other **modular shelving** (Q) to create an efficient germination space that can also be used to grow out the plants longer if there isn't a greenhouse to move them into. This shelving can accommodate **grow lights** (R), **germination domes** (S) over **1010** or **1020 seed-starting trays**, and **fans** (U). Heat mats (not visible in image) help with the germination of heat-loving crops, like peppers. **Pro growers** focused on microgreens will use shallow **microgreen trays** (T) and may opt for low-profile fans, like these **grow rack fans** (U), making for an efficient side-by-side grow rack layout. There are a variety of tray types shown here: **1020 bottom trays** that can be filled with **6-cell packs** (V) to make up a 72-cell tray that measures 10"×20" (thus "1020"), **1020 air-pruning propagation trays** (W), and **1020 deep mesh trays** (X) for soil blocking. Both 1010 and 1020 trays can be combined with either low- or high-profile humidity domes. Some germinates (like tomatoes) will be potted up into larger 2", 3", or 4" pots. The **colored pots** (Z) can be fun and helpful in sorting crops. (These colored pots are from Bootstrap Farmer.) We find them especially handy for growing out annuals and perennials.

Overall, I like biodegradable pots and soil blocking because they reduce my use of plastic, but air-pruning trays are heavy-duty and long-lasting, so they also cut down on plastic use. Paperpots are great for **pro growers** to maximize space and time for certain transplants. I think flimsy plastic cell packs are no longer applicable in the world today. Although they are still commonly what we find plants sold in at most big box nurseries, they are wasteful. Reusing any pot and tray you find is awesome—and yes, you need to clean them and disinfect for disease as needed and to your standards.

3. Garden Starts 85

Task Flow Setup for Garden Starts

FOCUS
DIY Greenhouse Projects

This potting setup (below left) at Juniper Farm has a 50-gallon barrel cut in half to hold soil medium; the space between the halves is where you set your trays for filling.

These seedling tables (below right) in my greenhouses have rebar legs for easy adjustment on uneven ground. Here, they have been moved above raised beds that have been cover-cropped.

FARM FEATURE
Potting Up at Hedgeview Farm Organics

Brenna runs an organic market garden with 2 acres of vegetables. In April, once her crops are germinated in open trays, she begins the potting-up task. The **PRO-MIX potting soil** (1) is broken up in a cattle trough set on cinder blocks and wetted down with a mister nozzle. The cell packs for this crop of **Salanova lettuce** (2) (a popular crop for her) are placed into their trays on her workbench. Potting mix is shoveled onto the **tray** (3) and the **potting up** begins (4). Her **salamander heater** can be seen ready for action (4); it will keep the small greenhouse warm on cold nights. The open

3. Garden Starts

trays are positioned to the right of the **cell trays** (5), making it easy to move back and forth with the little thread-stage Salanova **germinates** (6). **Popsicle sticks** with **waterproof marker labels** (7) are transferred to the new cell tray once the open tray is emptied of germinates. All the essentials are within this busy **"work corner"** (8) of the small greenhouse. Brenna's **widger** (9) is a small tool used to lift seedlings and move them into larger pots, and her design of choice is a cut-off piece of plastic fork! Tools need not be costly to be very effective. Once trays are full, she lets the seedlings settle in. The soil is already moist and the little transplants safe; she'll water them more later as needed. They grow out under the right temperature and humidity until time for hardening them off.

Brenna is third generation on her family farm; her children are the fourth generation. The farm has transitioned from a mixed farm run by her grandparents, to a crop farm and dairy run by her parents to her organic market garden serving CSAs and farmer's markets.

Potting Up at Hedgeview Farm Organics

TOOL TIME

Scaling-Up Your Starts

There are many ways to scale-up your garden starts for more productive transplanting. The photo below shows a great DIY scaling-up setup. **Cedar legs** (A) support an old **bathtub** (B) full of **potting mix** (C) ready to be **scooped** (D) into **cell trays** (F) supported by **bagel trays** (E) used as a cell tray station (they allow excess soil from cell filling to spill back into the tub when a **ruler** (G) scrapes the trays even). A **tray dibbler** (H) is handy for making impressions. The **seedling table** (I) has **built-in rebar posts** (J) to support hoops used with row cover for hardening-off. **Seed boxes** (K) keep seed organized by crop type. A **Kwik-klik drop seeder** (M) uses the right plate for the **dibbler** (H) for the **Paperpot tray** (F). A **tray** (N) keeps other essentials tidy and on hand. Other **tray models** (P) using Jiffy pots are on hand, as well as **germination domes** (Q) for smaller 1010 trays.

FOCUS
Popular Cell Tray Sizes

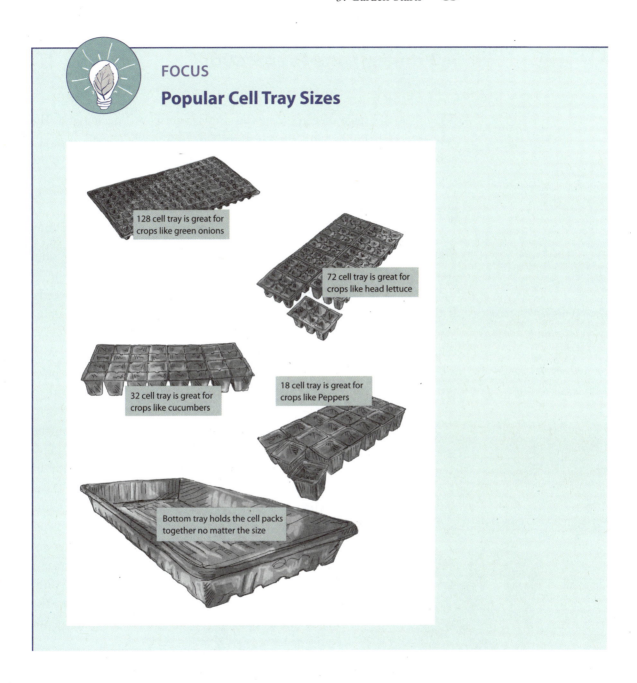

4. Primary Land Preparation

This is the stage of gardening you only do once—so do it right. The goal of **primary land preparation** is to completely get rid of weeds and amend the soil with bulk fertilizer and organic matter to improve the soil for future bed forming and years of crop productivity.

Preparing Land from Scratch

Start-up growers can use **poly plastic** or heavy-duty (commercial grade) **weed barrier** to fry the lawn or current ground cover in the shape of their new garden plot layout. Frying takes 2–6 months for most weeds. If there are still some weeds after frying for 6 months, it is best to cultivate, form the beds, and cover crop to further smother the most aggressive thick-rooted weeds, like thistles. Then growers can double-dig the soil with a **D-handle shovel** or **pick hoe** to break up hardpan and begin to aerate the soil. Compost is applied from a **wheelbarrow** with a **long-handled shovel**, and a cover crop is spread by hand in the fall to help the soil soften until bed preparation the following spring. **Scaling-up**, growers may use larger rolls of tarp and weed barrier and dump carts to spread compost on their larger plots. Handheld battery-operated tillers (such as those by Stihl) may be used to incorporate the compost and break up the turned soil. **Pro-up** growers will always lean toward renting equipment, hiring out custom work, or even purchasing a two-wheel tractor for their primary land preparation—if they are confident they will have further uses for it in the future. The **two-wheel tractor** and **rotary plow** are the superior equipment for smaller acreages.

The two-wheel tractor and rotary plow can do a great job plowing and turning over the sod before it's time to form new garden beds.

TOOL TIME
Wheel Measuring

Landscapers use **measuring wheels** often, but they are used much less frequently by small-scale growers. However, they provide an easy way of marking off land yet to be plowed for new garden plots. You simply reset the counter and then roll them along the ground in a relatively straight line to get your total feet between your layout stakes. Use smaller wheel units on flat land and larger wheels for grassy or rough terrain. Tapes can also be used to line up plots, and you should always use landscape flags and push-in and pound-in stakes.

92 *The Garden Tool Handboook*

TOOL VIEW: Two-Wheel Tractor

Two-wheel tractors can be fundamental to the primary earthworks stage and are worth borrowing, renting, or bartering for in these initial stages. Many growers that scale-up to become small-scale farmers and market gardeners (working about 1–3 acres) will transition or even start-up with two-wheel tractors.

My favorite two-wheel tractor implements are great for many growers and homesteaders. The **BCS 739** (A) is a practical, walk-behind tractor. It can connect at **the PTO** (B) to a number of useful implements: a two-stage **snow thrower** (C) for winter clearing of lanes and around greenhouses; a swivel **rotary plow** (D) for micro-plowing of flat and sloped land; a rear-tine tiller with **precision depth roller** (E) that can be used to break down plow furrows into plots and prepare new bed tops once the **power ridger** (F) is used to furrow paths and form bed shoulders. Once the beds are harvested, the debris can be flail mowed with a **flail mower** (G) that can also be used for clearing the edges of the garden plot and managing cover crops.

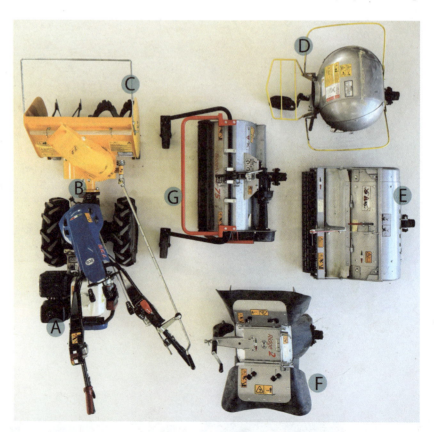

FOCUS
Tools or Tractors?

This book is primarily about **hand tools**, and it is possible to grow a garden and even a small market garden with *only* hand tools, as you will see in the pages to come. But if you **scale-up** or **pro-up**, you may want to add larger equipment. The **two-wheel tractor** is a popular next step. It's a lightweight, maneuverable, walk-behind tractor ideal for small-scale farming because it powers many of the implements commonly needed for garden tasks and property maintenance. This equipment is popular with growers who have more than 1 acre, especially if they are growing intensively as commercial producers or market growers. For a fuller discussion, you can check out my book, *The Two-Wheel Tractor Handbook*.

Primary Earthworks by Hand

Every soil can be opened using the right tool. This stage is called *primary earthworks*, but it is often referred to as *breaking ground* (and when the ground is very hard, this is literal!). When you break ground exclusively by hand, there are a few major considerations. First, equipment can help you make the straight lines you need to mark out your boundaries clearly. Be sure to use exact measurements and landscape flags. Next, you will need to understand your soil type to choose the best hoes to break the ground (See below). Also, many hands make the work light. This is 100% the time when you ask people over to pitch in. With enough help, you won't have to compromise on quality by reducing the depth worked or compost integrated. Lastly, when working by hand, integrate the bed-forming process into the primary earthworks by working up the soil and then forming your raised beds on the same day. This saves time, and you'll be able to see your progress, which will help you stay on track.

TOOL TIME
Hoe Design for Different Soil Types

There is an array of hoe forms that determine their function of earth working in different soil types. Many of these models are less known in the Americas than they deserve given the variety of soil types across the continent and the increasing popularity of small-scale farming, homesteading, and pro gardening. I have come across variations of these in Central America, Europe, and Asia, and am very pleased to see an increasing availability of these styles in the US, Canada, and elsewhere. I commonly refer to earth-working-hoes as **grub hoes** when they are narrower, and **broad hoes** when they are wider, but there are many variations. These great hoes are from SHW tools in Germany and available through Holden Tool Co. The diversity of these hoe blade's form for function has been put to the test in trials in my different soils on my land in Ontario. My farm has two major soil types: a loam soil, and a very pure clay soil. I also have plots with stonier and rockier versions of the loam soil, as well as gardens on sloping and flat land in Permabeds and terraces. All these different soils provided a great playground for tool testing hoe form for function.

When it comes to primary land preparation, choosing the right tool to suit your soil can make a world of difference, and a good heavy-duty hoe can be a great way to move and turn over soil by hand. Primary land preparation is well known to consist of **plowing** to turn over new ground, **disking** to chop the plow ridges, and **harrowing** with a S-tine to smooth out the disked land. Once this is done, the soil is ready for fine seedbed preparation and seeding. But we are mainly interested in hand tools in this book, and primary land preparation requires spades and hoes to take the place of the plow/disc and cultivators to replace the harrow. For hand tools the selection of the *right hoe and cultivator style for your soil type* and even the right weight (light or heavy-duty) is critical for your ergonomics and efficiency as a "human machine" doing the work by hand.

This **broad hoe** (1) will move a lot of soil with its wide and long hoe blade, especially in sandier soils where there is less resistance to its large form, allowing you to maximize its power to pull and move earth with its wide shape. This **spade hoe** (2) is a digging *machine*, cutting easily into a variety of soil types and can dig into stony loam soils much more easily with its point, and it can serve to trench out paths for Permabed reforming with ease. This **broad root hoe** (3) has a sharp, notched edge that helps when there are more roots present in the soil being turned over, like a weedy old field. This **oval hoe axe** (4) is a two-headed tool with an oval hoe that moves loamy and light stony soil nicely with less resistance than the broad hoe, and it has an axe on the other end for chopping through roots. This tool would be ideal for clearing and turning soil in a field overgrown in shrubs or even a forest being turned into garden. This **trident hoe** (5) has less resistance than a broad hoe in clay soils, allowing material to slide into the spacing between blades while also grabbing the material with its three blunt-tipped blades that act as mini shovels. This **clevis hoe** (6) has only two blunt blades for stony clay loam, where stones may get stuck between the trident hoe's blades. The **clevis pick** (7) is less useful in sandy soils, but it can break dense clay layers and excels in stonier and even rocky soils because the sharper points on the hoe blades break rock as you turn the soil.

Tools with multiple heads, or *duo-hoes*, can help you meet different needs as you work, like the previously mentioned oval hoe axe. Other examples include the **clevis pick/spade hoe** (8) and the **trident/border hoe** (9) with variations of the above-mentioned designs on one tool handle. But you will note that these last two hoes are actually a lighter duty tool. Any tool design can have lighter and heavy-duty versions (or weights). Heavier duty tools entail stronger handles, collars, and fasteners to go along with a heavier **hoe blade** (10). The hoe blade(s) go from narrower, shorter, and lower gauge (light-duty) to wider, longer, and higher gauge metal (heavier-duty); where the light-duty is best for hoeing weeds, medium duty for hilling crops like potatoes and finally more heavy-duty for primary earthworks. The same goes for **cultivators** (11), where the light-duty tools are used for cultivating weeds and mixing in fertility and heavier

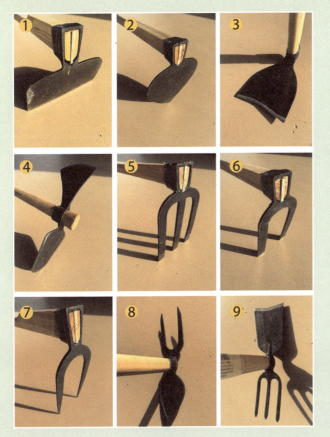

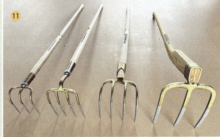

duty tools are used to harrow newly turned soils. You will also note the largest cultivator here has more of a curved axe handle designed for a chopping action into the soil!

An important principle to understand is that no matter the garden task you will find that for different soil types (loam, clay, sandy, stoney, rocky) you will be able to choose a tool weight that best suits. In this way a grower in a stonier soil may want a heavier gauge hoe for weeding than someone in sandier soil, and same for the hoe weight needed to do primary earthworks. One farm's hoe weight for hilling may be another's for primary earthworks!

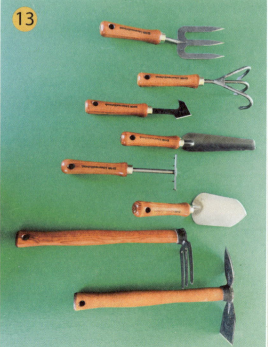

*Sometimes a heavy-duty grub hoe needs a **shorter handle** (12) to help turning soil in container beds or when working into a hillside where a long handle on a hoe will impede proper swinging into the slope. Many **garden tools** (13) have short handles and are designed for light-duty use, the handles are narrower, and gauge and weight of the metal is less than a short-handled heavy-duty hoe. Use tool appropriately for their intention and your soil type and context, and they'll last longer, be safer, and more ergonomic in hand.*

5. Plot and Permabed Forming

Like **primary land prep**, forming your new garden plots and Permabeds is only done once. Take the time to do it right. Many of the same hand tools used for primary earthworks can be employed in this stage to form plots and beds. After this stage, you may reform and reshape the beds seasonally, but you will never have to build them from scratch ever again!

Lay out the Plot

Now that the primary earthwork is finished, and you have all your site analysis data gathered, you can decide where to put your plots. The layout of a new garden plot is all about clarifying bed boundaries. First, you mark the primary corners of the bed(s) with flags. Use one of the straight lines you created during plot preparation to mark off the centers of all your *paths*—maybe there are 12. Then measure and flag the length of your *first path* and dig this material out (with a spade for harder unworked soil and a shovel for tilled soil); pile it to the right onto the soon-to-be raised bed. Do this for each of the 12 paths until you have 12 mounded beds. Rake them back and forth with a wide 30" garden rake to rough in the shoulders until they have firm bed tops. Define shoulders by measuring them and even squaring them with a string line or drywall square. By using a 30" garden rake you can rake the bed tops until they are just equal or a little wider than your rake, this is an easy, on-the-go measuring method. The paths should be of equal widths.

FOCUS
Permaplot Patterning

All growers need to decide the length of their beds and their width: this is their *architecture*. We also need to decide how many beds go into a plot, and, if integrating annuals and perennials, we must decide on a plot pattern, or ratio, of these different crops. This Permaplot has an organized pattern of three beds in perennial fruit, followed by nine beds of annuals and then another three beds in perennials. Three beds of annuals is a good unit for garden guild design, so plots of nine beds allow three of these "triads" for guild design and then perennial triads on either side.

This Permaplot has three beds in perennial fruit trees with non-spreading berry bushes and ground cover, followed by nine annual garden beds, and then three beds in raspberries (a spreading, perennial crop). This is the ratio of annuals to perennials I developed for my market garden over the last few decades. You can learn more about it in my book, The Permaculture Market Garden.

Permabed Architecture

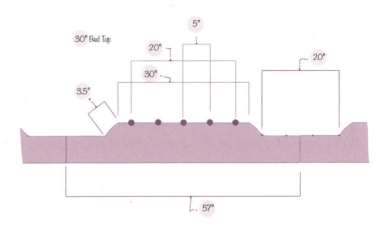

In order to make a good plot with great garden beds, you need to define the architecture of the beds. Here, Permabed architecture is showing a 30" bed top with room for up to five rows at 5" apart. The shoulder width is about 3.5" on either side, and a generous 20" path for easy access (and kneeling between rows) is specified. This is a great home garden and small-scale farm design. More extensive growers will want wider beds and paths. It is essential to know the width of your bed tops, shoulders, and paths—so you can manage your crops with the right tools. Do I want a wheel hoe blade that is wide enough to make 2 or 3 passes to keep my path clear? A wider rake to fit my bed top?

Bed Forming

A **start-up** grower can use a **tape measure** when forming beds, but as you **scale-up**, you will probably want to use a **measuring wheel** (A) to easily and accurately measure plot dimensions. Use **flags** (B) to mark bed boundaries and paths, and **push-in stakes** (C) to mark plot borders and perimeter alleys. **Pro-up** growers will make use of aerial imagery from drones to design their plots and blueprints to follow for plot layout and bed forming.

A 2-tine **clevis pick hoe** (D) can be used to break up the ground further if needed once the paths and beds are precisely placed. The 12"–20" wide paths, (30"–36" if using a Compost-a-Path method) can be dug and the loose soil applied to the bed top using a **long-handle shovel** (E) and pulled into the bed top using a **grub hoe** (F). The bed top can be roughed in using a **broad garden rake** (G) and then the bed top can be further subsoiled (the process of loosening soil down 12" to 18" to aerate and improve drainage in the soil). This can be done using either a **digging fork** (H) or (when scaling-up) a **broad fork** (I)—like this model by Meadow Creature that works well in heavy soils. **Pro-up** growers will usually make

Pro-Tip: *Once beds are formed and roughed in, they will often get a cover of composted manure and then be cover-cropped to get them through the winter without eroding. I use winter rye and red clover seeded 6–8 weeks before first frost.*

use of two-wheel tractors for this stage if they are growing over ½ to 1 acre of garden beds. Otherwise, even an acre+ of garden can be formed into a Permaplot garden with hand tools by a small crew of 5–6 people working together to make initial bed forming fast and efficient. If you have a two-wheel tractor, then the **rotary plow** is extremely maneuverable and can turn the top 6"–8" easily into organized rows. Making passes back and forth on either side of the *mid-line* of a future garden bed does a good job. Another method uses a power ridger to plow/till out the marked paths between two future garden beds and so partially form the beds on either side in one pass. See *The Two-Wheel Tractor Handbook* for complete systems using this equipment.

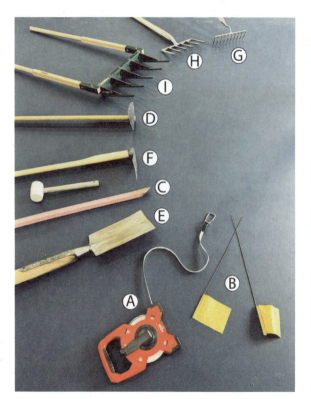

You can form new Permabeds by hand with many great hand tools, but a two-wheel tractor and power ridger can do a great job of turning the path material over and out onto the new bed tops on either side.

TOOL VIEW: Bed Forming

You will only form your plot and beds once using the Permabed system.

Caoba Farms, nestled on the edge of Antigua, Guatemala, runs an organic market garden. The well-formed garden beds are raised to help drainage in wet seasons and improve crop growth with reduced compaction.

(1) Here, plots are broken into eight equidistant earthen Permabeds for annuals between sections of three beds for perennials. (2) The land is prepared with a poly tarp to kill back weeds, and the soil with poor drainage is improved by using strawbales to form hügelkultur-style raised beds. (3) Permabeds are formed for a polyculture vineyard in this suburban site; weed barrier is used to zipper the beds to keep weeds out while establishing the new plot.

CASE STUDY
Container Gardening

Let's take a moment and turn to a different way of growing—in containers! Not all growing is done in the ground. Some growers—especially home growers—like to grow in container-style raised beds because they don't require bending over, and plants (and weeds) are within easy reach. Growing costs are much higher with containers, and this sometimes limits growers from larger garden productions they may otherwise enjoy. But benefits include ergonomics, aesthetics, and the ability to fit into an urban landscape, which is of great value to many growers.

Terminology: Terms like *raised bed* can be confusing. For market growers around the world, the term *raised bed* simply means a space of ground raised up a few inches by tilling or mounding with discs. A Permabed is a bed raised 4"–12" with very defined architecture that is never destroyed (the ground is never re-flattened). When we grow in beds sided with cedar boards, galvanized metal, and other non-soil material, we are growing in raised beds (of sorts), but this is really container growing.

Container Garden Beds and Plots

Container gardening is a specific way to grow vegetables, fruits, herbs, and flowers that involves planting in raised "contained" beds. Containers can be made of rot-resistant wood, like hemlock or these **cedar boards** (here, 2×10s), including two **long sides** (A), and two **short sides** (B) which can be stacked one atop the other and framed in with **wooden corners** (C) and screws, or **metal corners** (D), and supported along the sides with **2×4 or 2×6 wood boards** (E) or **metal flat stock** (F) to keep the boards from separating and forming gaps between them. With this system, you could build a 10"-, 20"- or 30"-high wooden bed that is 5 ft, 10 ft, 15 ft, or more in length at a standard 3 ft wide (about 32" of inside soil). **Cold frames** (G) can be made or fitted onto wooden raised beds. **Corrugated metal beds** (H) are another option—there are some interesting modular designs out there. Other options include **pots** (I) made of terracotta, poly, ceramic, etc.

5. Plot and Permabed Forming   103

These are great for urban growing and for decks, patios, and even potted production of crops like peppers, tomatoes, and high-value raspberries in greenhouses. **Grow bags** (J) are another lightweight option.

Note: The term *container garden bed* is more descriptive than *raised garden bed* because even a 2"–12" bed made of only soil is considered a *raised bed* in a market garden.

TOOL VIEW: Container Garden Beds and Pots

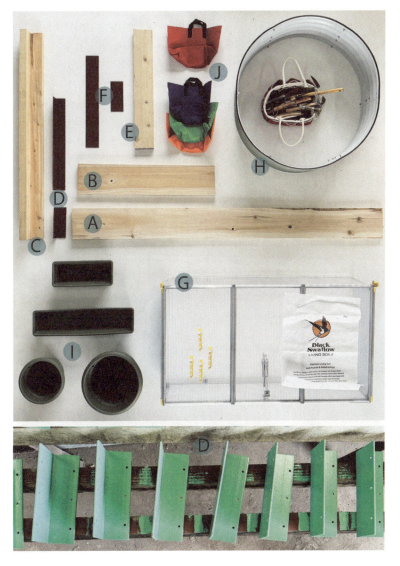

Container Shapes

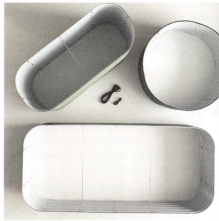

These Sproutboxes are a modular container garden system that can be put together in different lengths, widths, and shapes. Good application for urban yards.

Container Growing

Gardening in raised containers requires all the same tool types as flat-ground garden bed production but there is variation in the length of the tools. If your containers are low to the ground (6"–12"), most **long-handle** (~35" to 65"/90 cm to 165 cm) tools will work. These tools are typically used for flat-ground growing and are also very functional in short containers. The variation in length has more to do with tool type (spades and border forks are shorter than weeding hoes and rakes by ergonomic design). On the other hand, if your containers are high (16"–30"), then **medium-handle** (~16" to 30"/40 cm to 75 cm) tools are necessary. These tools are specially designed to allow you to reach across raised container beds—without the excess length that would make them hard to use. The one exception to this is the rake. Because you can rake along the *length* of a container garden bed, it is helpful to use a standard rake for seedbed preparation. Some garden tools for container growing are the same as for flat-ground growing (like the transplant trowel).

Short-handle (~6"/14 cm) tools like hand weeders and trowels are universally used in container growing and flat-ground garden bed production.

There are several more specific tools, especially for weeding, that are popular with container growers. The very nature of the container being of a material other than soil creates an edge between the garden bed and the container that needs specialized hoe knives that cut along the edges between the wood and metal siding and the soil. One of the most important tools/infrastructure in this type of growing is of course the container itself, so let's look at those options now.

The variety of container types in this outdoor space in Switzerland shows the many possible approaches to container gardening.

5. Plot and Permabed Forming

FOCUS
Designing Your Container Garden

There are many types of containers, and they should suit your site and needs. This series of **cedar boxes** (1) creates large planters for a rural homestead in Ontario. Cedar is used for its natural rot resistance. In this rural area, local white cedar is still relatively affordable (making it competitive against toxic pressure-treated wood and imported exotic woods). The tall planters have built-in trellises for tomatoes. The **moveable planters** (2) in this urban neighborhood in Philadelphia leave the appropriate clearance for pedestrians on the sidewalk. Some are designed to be moveable (with a pallet jack); others have no bottoms, and the concrete sidewalk was cut to allow perennial roots to penetrate deeper. The relatively cheap pine used for the **timber planters** here (3) has been treated with fire using the *Shou Sugi Ban* technique to make them more fire, water, and weather-resistant in a natural way. These planters, in a community garden in Dawson City, Yukon, are made with 3" square timbers and built using a log house, stacked approach. This makes for a very strong planter that will provide years of service. The uniform size and layout of these containers are sensible for a community garden where individuals pay for the annual use of a container. In other situations, like a homestead garden, it would be better to run the container beds as long as possible to facilitate easier planting, weeding, watering, and harvesting. A straight continuous container bed of 4' x 30' is much more efficient to manage than six beds at 5' long or three beds at 10' long.

Tools for Container Growing

Note, that the descriptions below are for the medium-handle tools (1); the short-handle versions are shown just below (2) with the same (A,B,C) references.

A **planting line set** (A) is used to run string to outline new beds and aid with marking rows for seeding. The **hand spade** (B) is great for double-digging in container beds to loosen the soil and mix in compost. A **mattock** (C) with a handle about 2.5' (78 cm) long can also be used to rework soil in spring to prepare it for new crops. A normal-length **shovel** (D) is needed to move soil from a wheelbarrow into a container bed. Similarly, a normal **garden rake with 6 tines** (E) is needed to make the nice, long passes along the full length of the container that are needed for seeding. A **smaller hand rake** (F) can be used for fine seedbed preparation in smaller containers. This **dibbler** (G) is just a tool handle without the blade; it's great for imprinting for bulb planting and leeks. This **hand transplant spade** (H) will easily cut through roots for planting and transplanting perennials. The **trowel** (I) is always useful for planting. The **cultivator** (J) is needed to mix in fertilizer and aerate between rows. **The halfmoon hoe** (K) is great for weeding between crop rows, and the **Cape Cod weeder** (L) is also a good short-handled hoe for this job. The **raised bed hoe** (M) is great for breaking up ground, hilling, and whacking weeds in low container beds. This **weeding fork** (N) is great for weeding and harvesting small roots in low container beds. This **dandelion trowel** (O) is used for digging up dandelions and roots in low container beds. A **Great Dixter fork** (P) works great for deep root crop harvest, and it can turn a bed like a subsoiler (as a broadfork does on larger plots). This **leaf rake** (Q) is used to clear off mulch from bulb crops in spring and clean up debris from bed tops; it can also double as a tine weeder for blind weeding small crops in low container beds. The **potato fork** (R) is a great addition for growers with more and larger containers holding more crops that need digging.

5. Plot and Permabed Forming 107

TOOL VIEW: Container Growing Tools

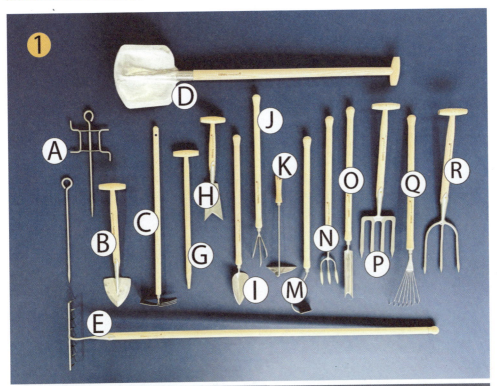

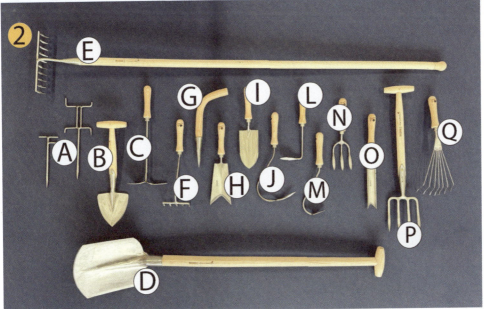

6. Fine Seedbed Preparation

This production stage is about creating a loose medium on your garden bed tops—one that is suitable for germinating seeds and planting transplants.

How Fine Is Fine?

A finished seedbed is often defined as a *fine seedbed* because the soil texture on the top is much finer than the roughed-in bed; it is fine enough to seed small-seeded crops like turnip, mizuna, carrot, and radish. You want the bed to have finer particulate sizes on the top of the bed to make it easier for seeders (and weeding tools) to move through the soil and to ensure the buried seeds have good seed-to-soil contact for improved germination. Large clods of dirt won't allow this, and debris and "trash" from crop residue will interfere as well.

On the other hand, if you are transplanting melons, zipping them in weed barrier, the bed could have more stones and debris without much effect on efficiency. Furthermore, the practice of crimping rather than mowing cover crops creates debris-rich beds that sort of change the meaning of fine seedbed to "finished" seedbed. Therefore, fine seedbed preparation is very much about understanding the crops and production models you are using and finishing your beds to suit those constraints. Seeding and transplanting requirements will change your goals about finishing your beds, as will the way you seed, weed, and harvest.

Typical Tasks

Typical tasks for finishing a seedbed include *tilling* and *raking* to make the bed loose but firm on top and deep and worked beneath. Some growers focus on subsoiling with broadforks and raking the surface. This is often followed with periods of pre-weeding before seeding crops and mulching before transplanting crops.

6. Fine Seedbed Preparation

FOCUS
View of Beds in Zurich

These beds in this **community garden** in Zurich, Switzerland (1), have been rough-formed in the spring and are ready for a rake or tilther to be used to turn them into a fine seedbed. Here, other **garden beds** (2) have been left over winter with crop debris mulch on the surface; they will need to have their form roughed in again and then a fine seedbed made for the spring planting. This roughing in reforms the architecture of the bed, the shoulder, and the height lost from seasonal settling; fine seedbed preparation follows to make it plant-ready.

Tools for Finishing Seedbeds

Start-up growers will often use a digging fork, **garden fork** (B), or a **broad hoe** (C) to turn and loosen bed material from the winter period of rest. **Scale-up** growers will use a **broadfork** (A) to cover more ground and subsoil the entire bed. A broadfork loosens the soil deeply, making it easier for transplanting with a trowel and for the roots of plants to reach down into the ground. **Small weeding rakes** (D) can be used to pick out stones and debris clusters ahead of using **medium rakes** (E) and **wide garden rakes** (F). All will be used to work the bed top perpendicular to the length of the bed. Continue to work the bed top and pull debris and stones into the path and break up earthen clods into a finer tilth on the bed top—making it good for seeding. **Pro-up** growers will often use a **seedbed bed roller** (G) to pack the soil and make the top firm. **Start-up** growers are more likely to do this with the back of the rake, raking the bed-top surface. A tool gaining popularity is the **tither** (H); it has a small tine tiller powered by a cordless drill and battery. This new model from Johnny's Selected Seeds has been updated for increased utility in more soil types. Overall, the tither performs best in improved garden soil in raised beds; it is not intended for primary tillage or for working heavy clay soils. ***Note:*** A concrete roller (see "Seed and Kill Your Cover Crops," below) can also be used to pack a bed top lightly ahead of seeding.

TOOL VIEW: Finished Seedbed Preparation

6. Fine Seedbed Preparation   111

Your garden beds will never need as much working of the soil as the first year you make them, but they will need to be loosened with a hoe or subsoiled with a broadfork most years before raking them out into a fine seedbed to plant again.

The Johnny's Selected Seeds seedbed roller is great for firming up a bed top; optional row marker tubes can be attached to the cage roller to depress the soil at chosen row spacing (like 10" between 3 rows, or 5" between 5 rows). See feature on row marking and dibbling, later.

Get Ahead of Weeds

Pre-weeding is often overlooked or underperformed because it requires time right in the middle of a busy garden season, but it is one of the most important garden tasks. **Pre-weeding**, or *false sowing*, is done by preparing a garden bed's tilth for seeding as a **stale bed**, then delaying seeding and instead allowing weed seeds to germinate. These weeds can then be removed while they are small and vulnerable. This period of pre-weeding is then followed with

Pro-Tip: Evaluate tool purchases by finding weak links in your operation cycle. Consider carrots, for example. They are slow to germinate, so the bed can get very weedy. Do you have a specific tool to help carrot weed management? Transplanting helps crops get ahead of weeds. However, research shows carrots don't transplant well. In fact, it is common practice to stale seedbed (an important market garden technique) by using a tarp, flame weeder, and/or tine weeder to pre-weed the bed before you even seed it to solve the weeding weak link!

seeding and transplanting into *a clean bed*. Soil is full of weed seeds in the **weed seed bank**, and pre-weeding should remove these from the top 1"–2" of bed top without bringing more up in the process.

Pre-weeding is especially useful for slow-germinating/slow-growing crops like carrots, onions, and parsnips because they are not competitive against fast-growing weeds. Timing is everything, and your pass with your tool of choice should occur after the weeds have germinated and, ideally, when they are in the **cotyledon stage** of growth. This often occurs **7–14 days** after the stale bed is made, depending on the conditions for weed growth (moisture, heat, etc.). A more intensive grower will want to maximize their garden beds and speed up the stale bedding process by using sprinklers and/or row covers to stimulate weed germination.

Pre-weeding Tools

Start-up growers can use their **wide garden rake** (A) or even a **collinear hoe** (B) to lightly cultivate the soil. Light raking/cultivating of the bed-top surface after the weeds have germinated uproots them, and they can be left there to desiccate in the sun. **Scaling-up**, growers may speed up the time needed to false sow and increase the quantity of weed seed germination by irrigating with a **sprinkler** (C) or by some other means increasing bed-top moisture; **row cover** (D) helps speed up the process by increasing bed-top heat. **Black plastic tarps** (not pictured) can be used to cover bed tops and fry the weeds.

Direct-seeded crops, like carrots, *require* pre-weeding, but transplants can be planted into weed barrier using the Zipperbed method or burning holes in weed barriers using a **torch** (E) with a plastic mulch hole **burner head** (F). These rolls of **weed barrier** (G) from Farmers Friend have the holes already burned. The Zipperbed method calls for two pieces of weed barrier lined up in the mid-row of a garden bed, with the crop planted where the two pieces meet. It's a great method for one-row-per-bed crops, like squash, and the rolls can be reused. **Scaling-up**, growers may consider

6. Fine Seedbed Preparation   113

using weed barrier and black tarps to cover and fry entire plots, or they may choose to use a **backpack flame weeder** (H). A great scale-up tool is a handheld **tine weeder** (I) which can flick up small, white-hair-stage weeds with minimal disturbance to the soil. Growers looking to **pro-up** can opt for more specialized, full-bed tools, like the **manual tine harrow hoe** (not pictured) from Terrateck, or a **multi-torch flame weeder** (J). It is also possible to use your cordless-drill-powered **tilther** (K) to make a shallow pass to pre-weed before seeding.

TOOL VIEW: Pre-weeding

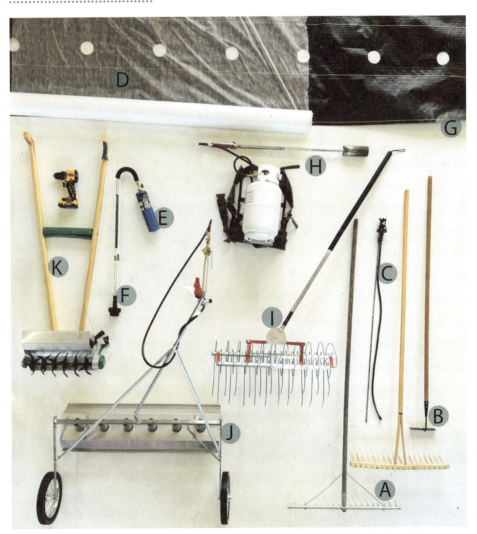

TOOL TIME
Flame Weeder

Flame weeders should be selected and adjusted to fit your bed-top width. You can choose the number of torches and the width of the flame weeder. The two-wheel model will keep the wheel tracks off your bed top. More propane tanks may be advisable for growers with longer beds, and the addition of wheels can be considered to stabilize the tool for longer passes. An offset handle makes this task more ergonomic. To give you a sense of how these work: the 5-torch unit from Flameweeders.com gets roughly 14,500 lineal feet of pre-weeding burn from a 20-pound propane tank.

One of the reasons **pro growers** will opt for a flame weeder is because it is possible to prepare a stale bed, wait a week, and seed your crop of carrots, and then in another 5–7 days you can pass with the flame weeder as a form of **pre-emergence weeding**, leaving the bed clear of weeds as the carrots emerge. This shortens the amount of time needed for pre-weeding since the crop seeds are already in the ground when the flaming occurs. With other, physical methods, like tine weeding, you can't do this. **A word to the wise:** If you are *even a day late* in flaming, and your carrots have already started to emerge, you are too late; you will kill your crop alongside the weeds.

Pro-Tip: If you put row cover or plexiglass over just 1–2 bed feet of your carrot plot, the seeds underneath will germinate first—showing you the right time to make your flame weeding pass because the rest of the crop will be a few days away from germination.

FOCUS
Succession Bed Preparation

Over the course of the growing season and from one season to the next, your garden beds will need to be re-prepared to ensure each new planting has a good, weed-free bed. Succession bed preparation is the series of tasks involved in re-preparing your garden beds to plant anew. This usually entails removing weeds or maintaining weed-free conditions, managing any cover crops, and preparing a new fine seedbed. The tasks for **succession bed preparation** change according to your crop plan, style of production, and scale. Some growers will "turn over" whole plots of beds in late spring; others will "prepare" one new bed at a time, as needed.

Techniques for re-preparing garden beds make use of many of the same tools needed for general garden bed preparation and fine seedbed preparation. Particular importance is given to turning over beds that are in crop or cover crop by killing the plants and turning them into the bed to help re-prepare the bed for a new crop. For instance, **tarps** can maintain weed-free conditions by frying weeds, crop residue, or cover crops and increasing decomposition to make room for a new crop. The cover crop and crop debris can be mowed using a handheld electric weed eater or scythe. This debris can then be buried for decomposition in the garden bed using a spade and shovel and turning path soil onto the bed top. It might be more efficient to simply cover the bed with that aforementioned tarp and let it fry and decompose for 1–3 weeks (depending on debris mass). These garden tasks will effectively turn cover crop and crop debris into a **green manure** to provide fertility to future crops. This is another stage where **pro-up** growers will find the jump to two-wheel tractors a powerful ally. The **flail mower** makes quick use of cover crops and crop debris, turning them into a green manure by incorporating them into the bed top with a **rear-tine tiller**. A **power ridger** on the two-wheel tractor is an awesome piece of equipment that can throw path soil onto both adjacent garden beds and cover over the flail-mowed debris allowing you to enhance decomposition with soil coverage without disturbing the bed soil ecosystem with the tiller. When you do till to re-prepare the bed top, you can do a shallow till of this recently applied path material (never disturbing the bed's core). Learn more about Permabed design and soil ecosystem health in *The Permaculture Market Garden*. Growers operating *with only hand tools* can manage cover crops by providing more time and space and making use of tarps to fry this back and add nutrition to the soil as they decompose under the cover.

7. Seeding and Planting

Seeding is the stage at which you will most likely want to have a new tool or two. A simple push seeder that costs a few hundred dollars will allow you to make your rows straight and get your seeds properly spaced and buried to the correct depth. Seeding and planting must be considered in relation to the tasks that go hand-in-hand with them, namely row marking and dibbling, because you need to make sure your rows of crop are easy to seed and (later) to weed and run irrigation on them too!

Organizing Rows for Seeding and Transplanting

Pro-Tip: *A bed marked with a tool is precise, so you can use more precise tools to weed those rows.*

Row marking is the act of leaving equidistant line impressions in the bed top's soil surface to mark where to seed rows of crops. **Dibbling** is similar but leaves small hole impressions to mark where transplants or seeds will be placed. Unlike row marking, dibbling marks both the *between-row* spacing and the *in-row* spacing. Another form of row marking, called *gridding*, leaves a graph-paper-like impression to help with both seeding and transplanting. Generally speaking, dibbling, row marking, and gridding are all referred to as "row marking." Refer to your **master crop data sheet** for best spacings for each crop before setting up your row markers. By marking **equidistant rows**, you will be able to seed straight along the length of your bed, and any future post-emergence weeding will be much easier to manage with hoe widths fit to your spacing.

7. Seeding and Planting **117**

FOCUS
Master Crop Data Sheet

A master crop data sheet is a spreadsheet that lists all the crops you grow in one column and adjacent columns have all the relevant data for design: days-to-maturity, rows per bed typically seeded or transplanted, pest associations, and other crop-specific information. You can draw much of this information from seed catalogs and other resources, and you can add new columns as your need grows for new crops and new information categories. For perennials systems, I have categories for "Ecosystem Layer," such as emergent tree, small tree, shrub, bush, larger herb, small herb, ground cover, vine, and root crop. This helps me organize the perennial plants I grow in my food forest into relevant categories based on their form.

For a **start-up** grower, using a wide garden rake with row markers slipped onto just some of the tines is the most affordable way to make rows or grids. **Scaling-up** growers will look to rolling row markers and dibblers to mark multi-rows with one pass.

Row Marking and Dibbling

Start-up growers may simply use a **hand dibble** (A) to break the soil surface and leave an impression to mark where to plant. Growers can measure the distance between dibble impressions with a simple **ruler** (B) or **tape measure** (C). This is standard practice in container gardening, and you can then come through and place your transplants with precise spacing (one per dibble impression). For flat ground and Permabeds, a **long-handled dibbler** (D) can be used to dibble or be dragged to mark a single row. You can also use a **wide garden rake** (E) with **row markers** (F). These are simply cut pieces of tubing that are placed over the tines at the correct distance. Dragging the rake straight and precisely down the bed will leave row marking at the desired spacing. This technique isn't 100% because the tool operator may zigzag a bit; however, the distance between markings is equidistant because row markers are a fixed distance apart on the rake tines. Arguably, this is what is most important because it allows you to select a weeding hoe with a width that will pass *between* future, equidistant crop rows. **Scaling-up** growers will use rolling row markers and dibblers that mark all the rows in one pass. Different-sized **dibbles** (I) can also be clipped onto the bed roller for use in soil or to pierce plastic mulch. Others may prefer a rolling row marker, or gridder style, that marks the rows while also making perpendicular markings across the bed top that can be used to organize in-row planting distances. Tools like the **VSS Aussie Gridder** (J) or the Johnny's Selected Seeds **Matrix Row Marker** do this.

My own row system requires a particular plant spacing, so I manufactured my own **rolling row marker** (K) that allows me to easily choose between planting 1, 2, 3, 5, or 7 rows without switching roller heads. You will notice that the rolling row marker has all seven marking discs on one tool head, meaning when I mark a new bed I can then choose which of the marked rows I want to plant (changing the **between-row spacing**) and which between-row marks I want to use for **in-row spacing** (space between the plants in the row). This means I can mark all the beds in a new plot area and *then* finalize the crop selection. For instance, I could plant just the middle row in squash at 24" in-row spacing in bed 1, and five rows at 5" spacing between rows and 5" in-row for spinach in bed 2, and two rows at 20" spacing between rows and 10" spacing in-row for kale in bed 3, etc..

7. Seeding and Planting

My method of having a continuous middle row and planting this middle row and its side rows as needed is called "123 Planting," where you choose to plant 1, 2, 3 or more rows to surround the center row, see more details below.) **Scaling-up**, growers may begin to use **rolling dibblers** for efficiency, especially the cost-effective **single-row dibbler** (L), like this model by Two Bad Cats, that can be run down any rows that have already been marked by a row marker. **Pro-up** growers will focus on multi-row and easily **adjustable dibblers** (M and N) that suit their crop row-spacing maximum. (*Note:* Leeks require a deeper dibble impression to help maintain a white stalk. Growers specializing in leeks or with larger-scale production may desire a specialized dibbler or leek puncher for this crop.) **Pro-up** growers may opt for a multi-row seeder (see "Seeding," below), a powerful tool that has enough weight to firm the bed top while placing the correct amount of seed in the row at equidistant spacing—all with one pass. But the cost/benefit needs to be understood, and the time needed for calibration limits its use to larger plot cash-cropping. A multi-row seeder could technically be used as a row marker for transplanting by using it without the seed hoppers.

TOOL VIEW: Row Marking and Dibbling

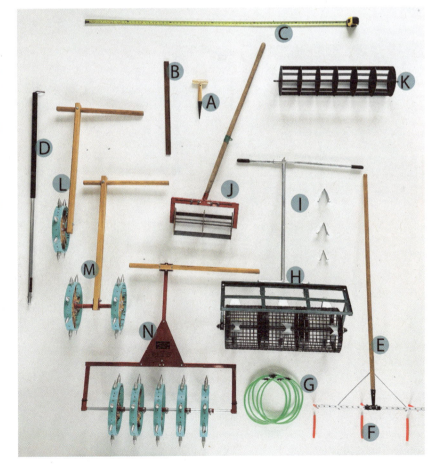

TOOL TIME
Original Tool!

The dibbler may very well be the original garden tool. A simple stake with a sharpened end could be used to open precise holes in the ground for seeds in early garden landscapes. What early growers didn't have was an easily *adjustable* dibbler. Modern dibblers are designed to dibble multiple rows at once and change **between-row spacing** and **in-row spacing** to suit **recommended crop spacing.**

Pro-Tip: *Pro growers with many different crops being transplanted will want a dibbler that is quick to adjust, or they may want to have a few dedicated dibblers.*

This leek dibbler has 15 cm (6") punches attached to the bottom, and you can see the three round row markers that align the punches for the next row to be dibbled and punched. Credit: Gemüsezeit Altluneberg GbR

FOCUS
123 Planting

The principle of **123 Planting** is important for row marking and may lead growers to desire a more versatile row marking system. In 123 Planting, there is a constant middle row, and then additional rows are added 5" apart, and you can choose to plant 1, 2, 3, or 5, or even 7 rows. This principle focuses on the concept that all crops have ideal rows per bed for a given bed width. Some crops will be best managed at 1 row, 2 rows, 3 rows, or more. If your row-marking tools create an odd number of row marks (like 5) and are centered on a single middle marker, then only one tool and one pass are needed to mark any bed. If your row-marking system has rows that aren't equidistant to a center row, then multiple tools are needed for row marking, and bed marking isn't consistent. See the homemade marking tool setup for bed marking using the 123 Planting principle. Using the 123 Planting Method, I grow either one crop in the middle row, two crops in the outer rows, or three crops in all three rows, or I split the difference between the rows for 5 crops. Simply seed or transplant into the rows needed: 1 inner row for squash, 2 outer rows for beans, 3 rows for kale, 5 rows for radish, etc. I never plant four rows, because that would require a unique row-marker set.

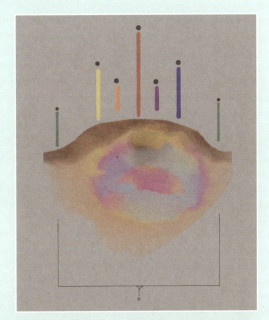

FOCUS

Calibrating Your Jang Seeder

The **Jang seeder** (A) can be calibrated to seed most any vegetable with great accuracy. Position the **back wheel** (B) between two **blocks** (C) and the **front wheel** (D) in any **wheel stabilizing device;** here, a tray works fine (E). Select a **seed roller** (F) based on the recommended roller for a particular crop. Seed rollers have different quantities and sizes of holes to allow for lifting **seeds** (G) like cilantro from the **seed hopper** (H) and depositing them through optional furrower types, like the **double-disc opener** (I) for heavier soils or the standard **furrower shoe** (J). These shoes are adjustable for different depths of deposit into the soil using the **screw tightener** (K). Once deposited, the seeds are covered by the **drag plate** (L) and packed by the back wheel for great soil-to-seed contact to improve germination.

Further calibration can be done by selecting different **sprockets** (M) for **positions** (N) and (O) with different quantities of **teeth** (like 10, 11, 13) so the **drive train** (P) turns the **roller** in position (Q) relative to the front wheel, which drives the chain movement. Choosing different **front** and **rear sprockets** will change the seed spacing in the soil. You can also adjust the **metering brush** (found inside the hopper and against the roller) to better hold the seeds against the turning roller for proper pick-up. To access the sprockets, lift the **protective cover** (R), where the **information sticker** (S) provides in-row seed deposit spacing based on hole quantity in seed rollers relative to the number of teeth in each front and rear sprocket.

To fine-tune your seeder calibration, set your seeder on the blocks and lift the **front**

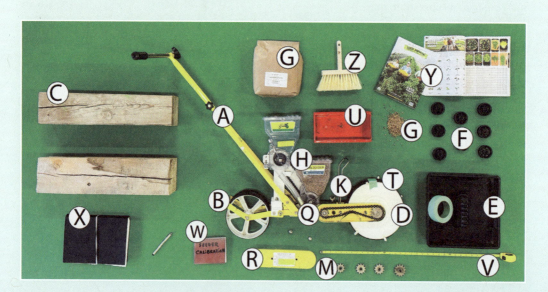

wheel (D) and mark it with **painter's tape** at location (T), turn the front wheel one full rotation, and **watch the seeds** (G) fall into the **seed catchment container** (U). Then count the seeds and divide by the circumference of the front wheel, which you can **measure** (V). Put this data into your **seeder calibration book** (W) and compare this to your **seasonal field notes** (X) about germination and recommended spacing from **seed catalogs** (Y). You can try different **seed rollers** (H) and change the **sprockets** (M) to see if this changes the seed count. Precision seeding is of particular importance when growing many rows of a crop and/or for crops that have lower germination rates or slower germination, both of which benefit from a more precise seeding tool and technique. Don't waste valuable seed. **Clean up** (Z) after calibration. *Note:* You can also see the side-dressing fertilizer hopper attached above the **seed hopper** (H). Also, once you have your seeder calibrated or a desired seeding density in the row, you can weigh your seed before and after seeding a full row (say 50 feet). This will give you the weight of seed needed for 50 feet using the preferred calibration setting. Now you can better plan seed order quantities. When you load the hopper for seeding always put 40-60% more seed in the hopper than required for seeding the row.

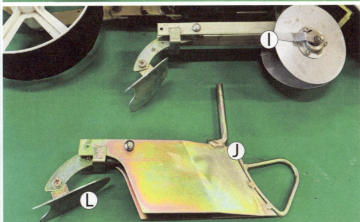

Furrowing and Hilling

An important variation on dibbling and row marking is *furrowing*. Furrowing is the act of leaving a trench (furrow)—or a series of them—to plant crops into. This is useful for crops that require a deeper planting, such as potatoes

**TOOL VIEW:
Hilling and Furrowing**

There are many tools to help you furrow. In fact, you can literally do this job by turning a collinear hoe on end and dragging it through the soil, but as you scale-up, dedicated tools will increase efficiency for crops like potatoes, leeks, beans, and more. Many pro growers will furrow the rows in the garden bed and lay Paperpot transplants by hand and then cover them over with tools like the Row Pro attachment from Johnny's Selected Seeds.

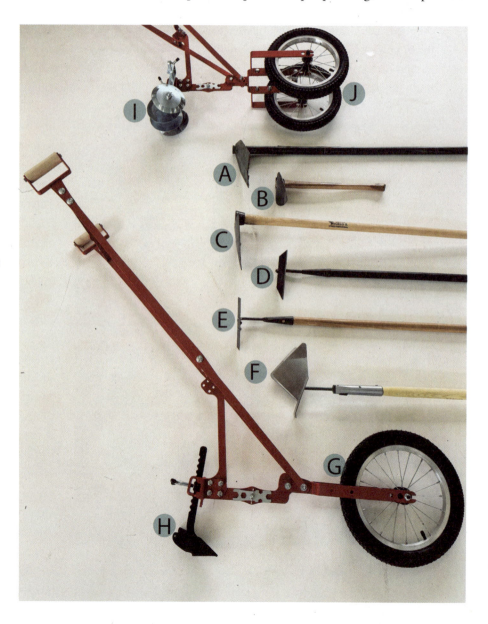

or leeks. Crops that require hilling often benefit from furrow planting because it leaves soil on either side of the crop row for future hilling. **Start-up** growers can simply use their **grub hoe** (A) to make a deep furrow for potatoes in loamy soil, or their **hand spade hoe** (B) for containers, or a **long-handle spade hoe** (C) for stony ground. Turning a **collinear hoe** or this collinear-style **"multi-hoe"** (E) on edge works fine to make small furrows for crops like beans. A **hilling hoe** (D) is lightweight and easy to move efficiently down rows for hilling the crops later. **Scaling-up**, growers may opt for a tool like the Johnny's **Row Pro** (F), designed to make a furrow *and* to fill it in. This allows furrowing to be used for most transplants by dropping the plants in the furrow and then pulling along to fill in the sides of the planting. This is especially used with Paperpot chains of transplants (without the Paperpot planter). **Pro-up** growers working more extensive crops will lean toward efficient systems like a **furrower** (H) attached to a **wheel hoe** (G) to make deeper, faster furrows for crops like potatoes, or a **multi-row furrower** to leave smaller furrows, as needed.

Field Transplanting

Transplanting is an important subject because it allows growers to get a 1–2 week jump on the season. Simply put, it involves starting plants indoors and then planting them out. But transplanting itself is much more labor intensive than seeding. Using tools to make this easier is helpful (See Tool View on page 127). **Field transplanting** is the process of bringing plants into the field and planting them into the bed top at the correct in- and between-row spacing. **Start-up** growers will simply use a **trowel** (A) to make a hole in a prepared bed to set the plants into and then fill in the hole around the base of the transplant. There are many unique trowels available—like this wood handle (with bark intact) **trowel** (B) made of reforged steel by Reforged Ironworks and Glaser's **right-angle trowel** (C), available at Johnny's Selected Seeds.

Transplants are grown in many sizes and shapes of trays and pots. **Start-up** growers will usually use small cell packs like these **reusable pots** (D) or **biodegradable pots** (E) for small transplants like lettuces and kales, and other, **larger, plastic pots** (F) for peppers and tomatoes. **Scaled-up** growers will lean toward reusable heavy-duty **poly cell trays** (I) or consider the

Paperpot system (R) or air-pruned trays like these from Farmers Friend (K).

Some growers will dibble instead of using a trowel for setting transplants; this is great for garlic, onions, and other small roots or bulbs. Two Bad Cats has a **long-handled dibbler** (G). When **scaling-up**, you might move to a multi-row dibbler, like this **two-row dibbler** (H) with adjustable dibble spacing; it's useful for marking holes in soil or roll-out poly mulch. If you use **weed barrier** (J) you will set transplants by burning holes or using the Zipperbed method. **Start-up** growers will be on their knees planting and need a **knee pad** (L), whereas the **scaling-up** grower will opt for other options, like a **hand transplanter** (M) and a **transplant caddy** (N) that allows you to *stand* and drop plants into the ground—with great ergonomic benefits. If you are working as a team, a **transplant bag kit** (O) can hold all your trowels and wooden label **stakes** (P). **Pro-up** growers will focus on perfect row spacings using tools like the **Matrix** (Q) row marker that creates equidistant spacing for transplanting rows and helps with efficient weed and irrigation management.

Pro growers will ultimately standardize everything to do with transplanting, from the way they start the plants, the trays they use, right through to planting into the field. The Paperpot system is a great example of this. With the Paperpot system, you can achieve not only precise between-row spacing by following your marked rows but also precise in-row spacing as the paper chains unfold the plants at exactly 2", 4", and 6" spacings. Most pro growers may want many of these tools to meet different needs: a Paperpot system for their closely spaced onions, a multi-row dibbler for garlic, and the hand transplanter and caddy as a means to transplant while standing for more widely spaced crops, like kale.

The Paperpot transplanter system saves time. But it also saves your body. Anybody who has ever bent over 30,000 onions transplants in April understands what I mean. For growers who want to grow a lot in a small space, it also allows you to get more **succession plantings** into the same space per season by growing crops as transplants, which you can, of course, do with other transplanting systems, but you would need to have many more human hours (people x hours working) to do this much hand transplanting. For the **pro grower**, this is a pretty practical and cost-effective tool. This tool wasn't available when I started market farming, but I have been

7. Seeding and Planting 127

enjoying it in recent years! Assembly is a breeze, which is great because I take it apart and reassemble it for landscape installation jobs to transplant herbs like thyme, too.

TOOL VIEW: Transplanting Tools

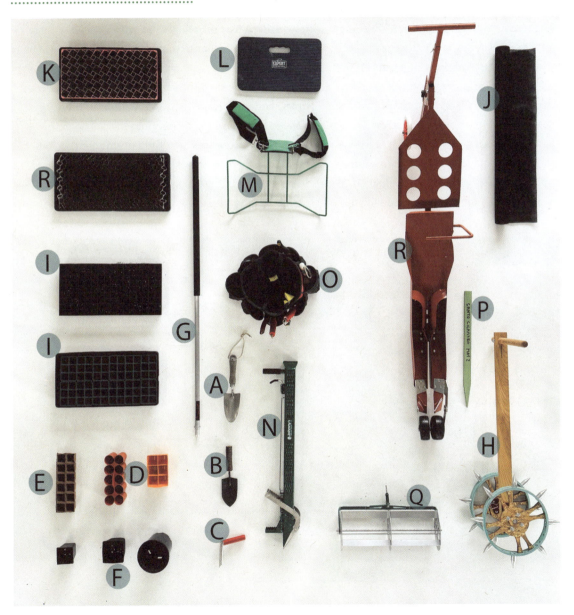

TOOL TIME

Paperpot Starts

The Paperpot system isn't for every crop. It is designed for crops with close spacings—including some crops you never thought to transplant, like radish! This system uses **paper-chain pots** (A) that come in 2", 4", and 6" spacings. Using an opening kit with **spreader bars** (B) and a **spreader frame** (C), the paper chains become an accordion-like **tray** (D) made of biodegradable paper honeycomb-like cells. The tray is placed into a **germination tray** (E) and filled with **soil** (F), leveling off the soil with a **wooden ruler** (G), at which point you remove the spreader bars. Now you can dibble the trays using a **tray dibbler** (H). Select your seed, such as this **bulk cilantro** from High Mowing Seeds (I), and use the **sizing tool** (J) to determine the right plastic **seeder plate** (K) for the **Kwik-klik drop seeder** (L). Hold this over the **dibbled tray** (M) and deposit seed precisely into the cell holes. Then use your popsicle sticks and **permanent marker** (N) to mark your tray variety and date and take notes in your greenhouse seeding **booklet** (O) before **cleaning up** (P).

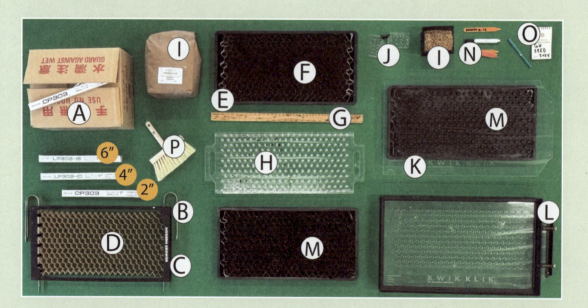

FOCUS

Field Tray System

I often bring transplant trays to the field stacked on bagel trays which can be placed in larger stacks on a cart. This means I can easily pull 6 to 10 transplant trays into the field and throw my transplant kit, row marker, and other gear into the cart, too. Because all trays are standard 1020 trays (meaning 10" wide by 20" long), a typical bagel tray can hold two trays with any cell count.

Seeding

Seeding is a critical stage for growers to get right; it is the process of getting seed into your bed at the correct spacing, depth, and soil contact. A *push seeder* can improve seeding efficiency, accuracy, and overall germination by dropping seeds from a hopper at the right interval into a trench furrowed at the correct depth and then packing the soil behind with a wheel.

Start-up growers would do well to invest in a push seeder like the **Earthway push seeder** (A) from the get-go. As soon as you seed **multiple crop varieties** (B) and more than a few bed feet of crops, it becomes obvious how worthwhile it would be to have a tool that could ensure your seeds are at the proper depth, spacing, and soil packing. As you **scale-up** on larger acreage, there are seeding tools that may be better for different crops, and as you increase acreage and require bulk amount of **seed** (C), it is helpful to fine-tune your crop production with specific seeding tools; for example, multi-row seeders like the **4-row pinpoint seeder** (D) are great for dense seeding of smaller and medium-sized seeds like lettuces and radish; they are also great for intensive beds of greens and baby carrots. Highly adjustable seeders like the **Jang JP-1 single-row seeder** (E) allow you to select from a wide array of seed rollers and gear ratios (see "Focus: Calibrating Your Jang Seeder," above) to create ideal spacing for any seed. This seeder has an optional **fertilizer hopper** (F), which is very handy.

Cover cropping can also benefit from a seeder. A simple **bag seeder/spreader** (G) is most commonly used for *broadcasting* cover crop seed but

Pro-Tip:
Seeding How To
Calibrate your seeder and load the hopper with 40-60% more seed than needed for the rows to be seeded. Walk slowly and methodically down the bed with the seeder lined up in your row markers. Check that the seed is not catching and sticking in the seed plate and that it is successfully dropping through into the furrow. Never seed when it is raining; always do it just before!

130 *The Garden Tool Handboook*

Pro-Tip: *Intensive seeders, like the 6-row seeder, work best in fine tilth soils (such as greenhouse beds). Also, any time row spacing is less than 4"–5" between rows, pre-weeding becomes very important because in-row and between-row weeding is difficult with such a tight row spacing.*

push seeders can also be used for this task. **Pro-up** growers will use many of these aforementioned seeders and also opt to get more specialized seeders specific to crop planting patterns like a **Johnny's six-row seeder** (J), which works great for intensive multi-row seeding, or a **Jang 5- or 6-row seeder** (H). Sometimes specialized and dedicated seeders for larger crops are desired, like the **Jang TD-1** (I). Larger-scale growers may wish to have dedicated seeders for certain cash crops to avoid having to make too many

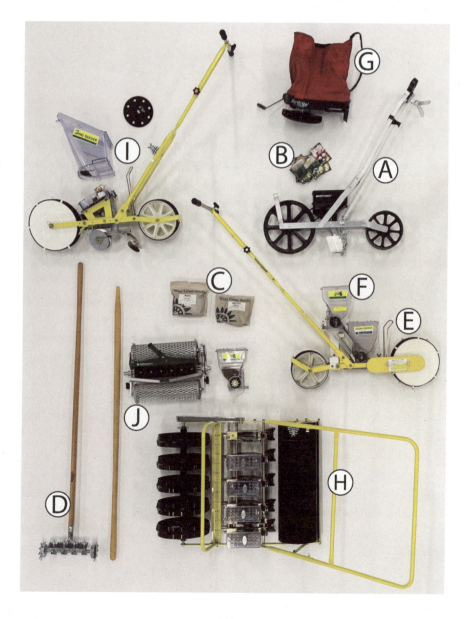

7. Seeding and Planting 131

adjustments and calibrations for crops they need to seed at regular successions in small windows of time.

Some growers have a **handheld seeding kit** (J) to carry **seed** (K) and other essentials like **Earthway seed plates** (L), or **Jang seed rollers** (M) and **sprockets** (N), a seeder **calibration record book** (O), **row marking tubes** (P) for your garden rake, **crop row labels** (Q), **permanent markers** (R), a **wheel measuring tool** (S) to measure off lengths of rows to seed, **scissors** (T) to cut open seed packs, and a good **transplanting trowel** (U) is also nice to keep in this kit.

TOOL VIEW: Handheld Seeding Kit

TOOL VIEW: Seeders

Pouring seed out of the seed hopper and back into the packet is best done with care. You can do this by squeezing the seed packet open in one hand and tipping the hopper carefully with the other. I do this here with a bit more flair—for the fun of seeing the seed falling; usually, the hopper lip is touching the seed packet. Another trick is to use a funnel!

The Earthway seeder is an affordable push seeder with easy-to-use, interchangeable seed plates for all common crops—arugula, radish, carrot, lettuce, peas, beans, corn, etc.

7. Seeding and Planting   133

TOOL TIME
Calibrating Your Seeder

Here, a 6-row Jang seeder is set up to seed only 5 equidistant rows. A multi-row seeder requires no row marking and will be beneficial to growers who seed longer (100', 200', or 300') beds of a single crop. The downside of these larger seeders include needing a lot of seed to put in the hoppers and all of the calibration needed for each seeder in this unit. For these reasons it is best suited for crops that are planted in large quantities routinely throughout the growing season for growers on a larger acreage scale. For instance, if you are seeding several beds of golden beets every few weeks. For any small rows of crops, I would always use a single-row seeder as it is faster to make adjustments.

FOCUS
Field Labeling

The pro grower will usually use larger stakes for field labeling to make reuse possible and ensure field data isn't lost from sun, weather, and abrasive activity. Smaller popsicle stakes used in the transplant tray really shouldn't be used in the field—they fade, break, and get lost in the soil. I have pallet loads of these stakes made for me at a local saw mill and use them on the farm and for landscaping jobs.

8. Irrigation

Irrigation is everything to do with how water is sourced, moved, and distributed to your vegetables. Growers have very little they can completely control in a garden environment. The addition of water when it's needed is one of them. Irrigation is not just helpful in a drought; it is used to prepare stale seedbeds, help germinate newly seeded crops (and so keep their succession timing correct), and encourage general crop growth over the season. However, water management is one of the first weak links to emerge.

Start-up growers will rely on hoses, over-the-counter sprinklers, and box store drip and soaker hose systems. They will often source their water from the household or farm supply/well and access it from a normal spigot. (***Note:*** *Quick connects* are great for switching between hoses when running water this way through your property.) **Scaling-up**, growers will look toward professional sprinkler systems like **mini wobblers** and full-plot drip systems that incorporate pressure regulators to make sure moisture is delivered evenly. As you **scale-up**, you will often source water from larger reservoirs such as **poly tanks** (sometimes connected to gutters/eavestroughs) or even ponds. Using these larger water sources requires investment in **gas or electric pumps** and the use of **filters** to make sure your drip systems don't clog. A homemade **manifold** can help keep hoses organized for water distribution. As discussed in the DIY section earlier in this book, DIY systems for **rolling up drip tape** are popular because drip tape can be reused. **Pro-up** growers will make use of larger-diameter **transfer pipes** and enjoy the functionality of **multi-connectors** that rise from the buried transport lines. Filtration will often be used for drip systems both at the pump and at the header hose to be sure no contamination clogs lines. The use of **heavy-duty drip lines** (or **Blueline**) is also worth the investment because they can be left in the field and used for a decade without getting damaged. Pro growers will often have custom-made irrigation systems for various applications such as overhead watering in a moveable hoop house, a misting system for their transplant greenhouse, and heavy-duty drip (Blueline) irrigation for their high tunnel.

Storage, Pumps, Manifolds, and Distribution

Irrigation comes in several parts; it is, indeed, one of the most complex aspects of small-scale farming when you consider the number of connections and options available. Irrigation is most easily understood by following the flow of the water: collect, store, pump, transport, distribute, regulate, irrigate, and conserve. For instance, on my own farm, I collect water naturally in a groundwater-fed pond, where it is stored. I then pump through transport lines to my main fields; there, I distribute the water via smaller transport lines to the header hoses that pass along the heads of my garden plots. Then the water goes through pressure regulators to ensure the thin drip lines don't burst and to help regulate even flow of water into all the drip lines (maybe 1–3 per bed of vegetables). Then it is about conserving the water with light cultivation, mulch, crop diversity, and healthy soil structure! I didn't mention filtration here, but it is a critical component of your system. There are several places where filtration occurs. First, if you are pumping directly from a deep-drilled well or municipal water source to your garden using household water, you won't need much filtration as this water is relatively clean and may already be filtered in the house. In this case, you want a filter at the start of a new header hose (right before the pressure regulator) to make sure no sediment clogs the drip tape or sprinklers. When pumping from a pond, many spots need a filter to keep your system clean (the same goes for roof catchment water systems). Let's now run through all the bits and pieces of a full system. ***Note:*** The sizes of piping in the infographic reflect different scales, not one whole system.

Water's Journey: Pumping to Distribution

Let's say your water starts in the pond or a storage tank. From here, it is pumped with either a **gas** or **electric pump** (A) that draws water through a **coarse filter** (B) in the **intake hose** (C) which is kept suspended in the pond (and not sucking in the mud at the bottom) by tying **ropes** (D) to the intake head and pulling them to a desired tension by attaching them to **T-bar** (E) in the banks to ensure the head doesn't sink into the bottom of the pond. **Camlocks** (F) are common quick releases for professional irrigation fittings.

A typical gas pump (A) will have a **pump inlet** (1), a **water filler plug** and **pressure gauge** (2), an engine with an **ignition switch** and **recoil start** (3), with a **gas tank** above it (4), an **air filter** (5), **muffler** (6), **overflow with gate valve** (7), and a **water pump outlet** with gate valve (8). Preferably, all this is housed in a sturdy frame, making it easy to carry.

From the pond or tank, water travels through a high-pressure **connection line** (G) to the main **system filter** (H) with its major components: intake with **pressure gauge** (1), **filter** (2), **clean-out pipe** (3), **filter scrubber crank** (4), outflow with **pressure gauge**, (5) and a sturdy stand. Similar to the pump, camlocks with a **male** (6) and a **female** (7) end help you easily connect and disconnect major components.

The water will then travel either above ground in transport lines or below ground in buried poly or HDPE lines. For above ground, I like lay-flat hoses that "lay flat" except when filled with water because it is easy to see where repairs are needed and to make adjustments. When you have a very good sense of your permanent plot layout, I suggest installing buried poly or HDPE lines. Smaller growers will use **1" poly water lines** (I), whereas larger projects may need 1.5", 2", or 3" **HDPE lines** (J). These lines will need various **fittings** (K) to connect them to the filter and pump assembly. They will run underground from there and make use of **couplers** (L) to connect two straight sections or other multi-direction **fittings** (M) and **end caps** (N), all of which are connected with heavy-duty **hose clamps** (O) of the right diameter. Slide the clamps over the hose or pipe, heat up the inside of the pipe with a **torch** (P), slide the fitting into the pipe, and then slide your clamps over the hose with the fitting inside and use a **socket wrench** (Q) to tighten. A socket wrench is a critical part of the tools that make up an irrigation kit because it is needed to help put all the pipes, fittings, and clamps together. (See sidebar "Irrigation Kit" nearer to the beginning of this book, under "Tool System Design").

The pipes can be fit with a **saddle fitting** (R) to send water up through a vertical pipe to your garden. Based on site needs, I make these 2-outlet and 4-outlet **multi-connectors** (S) with shut-off valves, camlocks, and hose quick connects. If using household water and/or distributing through **smaller water lines** (I) for smaller plots, then a normal **frost hydrant** (T) can be used as your main water **connection between transport lines** (J) and (I) and distribution lines to sprinklers and drip (see next feature). I

8. Irrigation   137

also build multi-connector manifolds using brass fittings or poly (don't use iron; it will rust fast!) for smaller plot water organizing. These **manifolds** (U) can have different types and quantities of **shut-off valves** (V) in practical arrangements for turning water on and off to distribution lines (which may be simple garden hoses from here on at this scale). In all cases, wrap **PTFE tape** (W) around fitting threads to prevent leaks, and make use of your other essential **hand tools** (Q) in your hand **tool kit bag** (X) to get the job right.

TOOL VIEW: Pumping and Distribution

A Journey of Water: Distribution to Irrigation

Once the water leaves the **multi-connector** (A) or other **manifold** (B), it can travel by a flexible and durable line. (I call this line a **"flex hose"** [C]) that can easily make the connection between your multi-connector (water source) to what ever line you use to water the plots or beds. This flex hose has a **female camlock** (D) connection to multi-connector and **male camlock** (E) connecting the **filter** (F) and **pressure regulator** (G). For smaller-scale plots, you want to just use a **Y-connector** (H) with an **anti-kink device** (I) attached to the **garden hose** (J), which can serve as the flex hose connecting to an irrigation system. Sometimes filters aren't needed in the field, and a **pressure regulator** is a stand-alone unit (K). Sometimes **larger filters** (L) are used to handle a greater flow of water to the irrigation system. **Start-up** growers may use a variety of **timers** (M), and scale-up with **electric valves** (N) and electric **controllers** (O) to help manage water needs. Many sprinklers may be used, including over-the-counter sprinklers, like **oscillating sprinklers** (P), impact sprinklers, and even this **pullable sprinkler** (Q) that can be pulled back along a path to water the garden beds on either side. **Scaling-up** growers will look for more specialized systems that have a more modular quality with multiple in-line sprinkler units, like the **Dan sprinklers** (R) and the **Xcel-wobblers** (S) that have different nozzle options to manage spray form and function. You can keep a collection of these in a **kit** (T). A sprinkler can be mounted on a **self-standing base** (U) or attached to a **metal spike** (V). Other sprinklers can be attached to the cross members of a greenhouse for **overhead irrigation** (W). Production in pots makes use of **drip tape** (X), and some growers simply need a **soaker hose** (Y) for small areas of edible landscaping, or they may just water by hand with the rain shower setting on a **spray gun** (Z), though specialized nozzles are often used in the pro greenhouse for transplant watering.

    A drip system has the following components in the field: input with a hose quick connect or **camlock quick connect** (1); usually a **PVC elbow** (2) to orient the water flow perpendicular to a garden plot along the heads of the beds; the **filter assembly** (3), including a cleanable mesh **filter cylinder** inside (4) and flush valve at the end (5); a **pressure gauge** (6); a **pressure regulator** (7) to maintain proper pressure for the irrigation lines

8. Irrigation   139

across the entire plot; output fittings like a **barb** (8) to insert into the oval hose, or **header hose** (9) for **drip irrigation** (10) or a **header hose** (11) for heavy-duty **blue line irrigation** (12) of perennials; **drip fittings** (13) are inserted into the oval hose using the tool kit's blue **hole punches** (16), whereas perennial irrigation makes use of longer-lasting connections with **T-fittings** (14) with **threaded shut-off valves** (15), which are easily installed with the tool kit's **red tube cutter** (16), hose clamps, and **Teflon tape** (17).

TOOL VIEW: Distribution to Irrigation

140   *The Garden Tool Handboook*

This pond was put in around 2011, and here it is on a fine spring day in 2017. From source to manifold to field irrigation, the system is about moving water from the pond to the crops.

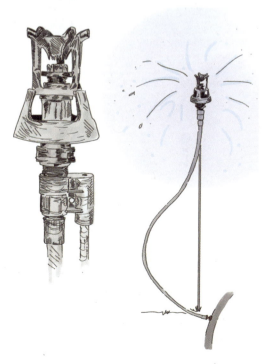

A Journey of Water from Manifold to Garden Row

Water may start at a pond or well, but when it comes to your immediate use and the organization of your irrigation, the story starts at the manifold. **Home gardeners** can use ¾" brass or poly manifolds; **homesteaders** may consider some ¾" manifolds for hoses and 1" for distribution within the plot. **Small and medium-scale farmers** may consider maintaining even larger diameters (1.5" or 2") at the manifold for further in-field distribution of water. For years, I distributed water between fields with a used 1.5" fire hose (cheap, strong, light, swivel connections = great).

FOCUS
When to Sprinkle and When to Use Drip

Every situation calls for its own type of water management. Sprinklers are best for stale seedbed preparation and for germinating new crop successions. If your newly prepared stale seedbed doesn't receive water, it won't germinate weeds on time and will take too long to pre-weed. Similarly, if a new crop is seeded and there is no precipitation that week, your succession schedule will fall behind. For instance, if you are planting baby carrot successions every 2–3 weeks and your crops are irrigated with a wobbler the day they are seeded, your schedule will remain precise because the stimulation for germination will occur immediately if moisture is maintained that first week after seeding.

There is a number of reasons sprinklers are used to maintain direct-seeded crops (like lettuces, carrots, and radish): (1) they benefit from full bed moisture to germinate; (2) they are in the ground too short a period of time to make it worthwhile to invest in laying drip tape; (3) it easy to move the sprinkler system between plots and beds when you need to make weed sweeps; and (4) the frequent weeding needed to maintain these types of crops is hard on drip tape.

FOCUS

Irrigation System

Irrigation systems should be designed to be crop-suitable and enterprise-specific. Some photos from my farm tell this story. This **field** (1) is planted into **winter squash** (A) which has its own drip irrigation set up under the plant canopy. The **secondary crop of lettuce** (B) is grown between crops of spreading cucurbits to make use of the **space** (C) before the squash canopies of adjacent beds meet. This line of wobbler sprinklers is helping the squash establish, maturing the lettuces and allowing a second or third cut and stimulating weed growth for pre-weeding of the **path** (C). The modular system allows the sprinklers (D) to be detached from the **distribution line** (E) and easily rolled into an adjacent plot before the squash creeps over the well-weeded and already harvested middle bed.

Pro growers will opt for versatile irrigation systems like the moveable full-plot wobbler system. These systems can easily be moved aside for cultivation and harvest and shifted to new plots for new bed germination and crop watering.

8. Irrigation   143

This **hoop house** (2) grows a crop of early greens (F) using **overhead sprinklers** (G) attached to hoop purlins (H). This design is effective because it allows this 50" **moveable greenhouse** (I) to slide off from the greens once they are established and onto a new plot for producing summer crops, *carrying its sprinkler with it*. This **permabed** (3) of **tomatoes** (J), makes use of heavy-duty **Blueline drip** (K) because it will be intercropped with young fruit trees, and the tomatoes are a living cover crop for a bed destined for future perennial use.

At the Edible Biodiversity Conservation Area, we have a wall of fittings organization bins and tools for the many irrigation works we need to manage 100 acres of water pumping, storage, distribution, and irrigation. A smaller irrigation storage system using modular bins, drawers, and a workbench could be useful for home, homestead, and small farms.

9. Garden Crop Maintenance

Garden crop maintenance is a catch-all term for a season-long production stage. It includes many of the tasks that are done routinely throughout the garden season, like suckering and trellising (though not all growers need to do those two things). An *extensive* approach (using wider row spacing and focusing on fewer crops for larger-quantity yields) is common for some market growers, and they may focus on a few crops that require neither trellising nor suckering: bush beans, pumpkins, and carrots, for example. For other growers, suckering and trellising are both essential tasks for crop maintenance. This production stage also includes important tasks that are spread out through the seasons, namely, pest and fertility management. Although I touch on these in other production stages, they are covered below in more detail.

Trellis Crops and Support

Crops that need support include some varieties of peas and beans, perennials like grapes and hardy kiwi, and many flowering plants. Tomatoes also need support. Popular options include round **tomato cages** (A) and **square trellis cages** (B). **Rebar** (C) makes an easy support stake for individual plants. I have also used electric fence **tread-in type posts** (D) for quick trellis support for overladen peppers. Wooden 4' and 8' **posts** (E) work for support as well—the smaller ones for shrubs, and the larger ones for trees—and you can write on them with paint markers before you get around to attaching metal plant labels. The smaller posts can be pounded in with a mallet or a **post-pounder** (F). If you are installing larger wooden posts, you will benefit from an **earth auger** (G), especially for 2"– 8" cedar posts (which are great for permanent perennial trellising). High-tensile **wire** (H) and **ground anchors** (I) are used to create strong wire lines for trellising crops like grapes using a **fencing tensioner** (not pictured) or an **in-line wire tightener** (J). **Lighter wire gauges** (K) are useful for making ties between trellis elements. **Tape guns** (L) can be used to attach tomatoes to their support string, or you can use **trellis clips** (M) and **stem hooks** (N) to support ripe tomato clusters—all of these fit nicely in a conveniently

9. Garden Crop Maintenance   145

worn hip **trellis clip bag** (O). **Start-up** growers often use cut-off pieces of hose and low-gauge wire for supporting trees; **scaling-up**, **tree straps** (P) can be very handy, alongside **binding tube** (Q) carried in a hip-mounted bag. Commonly used supports for annuals include **trellis netting** (R), **jute twine** (S), **baler twine** (T), or even **cotton twine** (U) tied to a top bar. In the lower photo, **trellis mesh** (1) is supported by the **top bar** (2) inserted into **PVC Tees** (3) on top of **rebar** (4) pounded in at the center of the bed top. If this infrastructure is employed in beds where weeds are managed properly, it can be semi-permanent. Crops like peas, tomatoes, cucumbers, etc., can be rotated into these "trellis beds." Always carry with you some **cutting tools** (V) for your strings and some **snips** (W) so you can sucker and prune.

TOOL VIEW: Trellising

Left: *I am starting to basket weave (aka "Florida weave") this bed. Basket weaving is an easy cost-effective way to trellis crops like tomatoes.*

Right: *Here, trellis strings tied to the wood support header are also anchored into the ground beside the crop with a metal spike or attached to wooden framing of the raised bed (as shown here). The crop is trained clockwise around the string as it grows.*

FARM FEATURE
Trellis at Beaverland Farms

Brittney Rooney and Kieran Neal have run **Beaverland Farms** (1) for 8 years. It's a 3.5-acre urban market garden built up across multiple properties in Brightmoor, Detroit. They sell mostly through CSAs, but they also have an on-farm market and take wholesale orders. They have many **high tunnels** (A), including these custom-built 20' x 60' tunnels made from 1⅜" top rail they bent themselves using a DIY plywood jig. These tunnels are set up to be ventilated with roll-up sides and end wall doors, and each has insect netting on the sides to keep out the bugs (cucumber beetles, especially). **Shade cloth** (B) is used for summer salad production in the cooler and cleaner shaded tunnel environment (with roll-up sides up). Inside the **tunnel** (2) is a busy space, and a DIY tool storage using **French cleats** (C) allows them to hang their essential greenhouse tools; milk crates using the same cleat system hold supplies like twine, trellis hooks, and irrigation fittings. When trellising, the whole crate of trellis supplies unhooks off the wall and is taken into the rows; it is then returned when trellising is done for the day—so everything stays neat. The trellis system for **greenhouse crops** (3) uses scrap **pipe** (D) resting free on top of the greenhouse purlins to hold the **Qlipr system** (E) hooks that are lowered as crops grow to keep them within reach for harvest and management. Beaverland Farms grow a lot of tomatoes using this system; the **clamps** (F) connect to the stem of the tomatoes, and the trellis line supports upward plant growth. This system makes it fast and efficient to lower plants as they grow and reclip the leaders to keep the plants at desired working heights. End-of-season cleanup is fast: you just unclip the two clips per plant and let the debris fall, to be raked up later.

Beaverland Farms also grows peppers in the **tunnels** (5); they find the best method is to prune the peppers to two main leaders in a V-shape (and in line with the overhead support) for easy trellising, and peppers are simply twisted around support lines as they grow, using clips if necessary. **Cucumbers** (6) are another of their excellent tunnel crops; they produce great-quality fruit (H) with their good production practices that include using the right fertilizer, insect netting to keep bugs out, and good management. As always in tunnels, trellising is important for this long-season crop that puts on a lot of growth and needs support to facilitate efficient harvest over the course of many months. Here, a support string is tied in the purlins and run to the ground from the get-go, and the cucumber vine is twisted up it as it grows. Initially, every other fruit set and all the suckers (which would become new side vines) are removed; once the cucumbers reach the top bars, they get "umbrella pruned" by topping the plant, causing extra downward growth that is nicely supported by the top bar.

Tunnel crops (7) need to be checked on regularly, and Britney and her team are on the job, checking 2–3 times per day to make sure tunnels aren't overheating (needing ventilation) or needing irrigation and that pests haven't gotten in (which would require netting repair). *Out of sight, out of mind* is an expression that should be held close for the greenhouse and tunnel grower. It is bad news not to check up on high-value crops in these high-value spaces. There is a lot of potential for things to go wrong, like over-heating and killing crops on hot days. All these tunnels have wax-operated vents that vent automatically, but roll-up sides need to be rolled up manually on hot days. Irrigation is on timers, but adjustments need to be made for seasonal extremes. On cold days and nights, the sides go down, and the doors have to be closed to keep frost-sensitive crops happy and producing.

In-Season Fertility and Pest Management

Fertility and pest management go hand-in-hand when it comes to the garden operation cycle. Good plant nutrition is an important part of pest management (unhealthy crops are susceptible to pests), and pest and fertility management use similar tools, like sprayers.

Growers use many methods for applying fertilizer. *Side-dressing* is a technique of applying fertilizer or other amendments *along* the row of a growing crop, often in tandem with cultivation to help work it into the soil and become soluble. *Top-dressing* is done by broadcasting the fertilizer over the entire bed top. (**Word to the wise:** Granular fertilizers high in nitrogen can *burn* or *kill* plants if too much is applied. Make sure you understand the correct quantities.) I manage fertility simply by dressing the bed or row by hand using a **glove** and a **container**.

A handheld sprayer can be used to apply foliar nutrients and soil soaks, like fish emulsion, compost teas, etc. **Scaling-up**, growers may opt for a **backpack sprayer** to make fertilizer application fast and efficient. **Pro-up** growers can add a fertilizer hopper to their Jang JP-1 seeder to side-dress with nutrients when seeding or by running it alongside a row after the plants put on some growth. Multiple backpack sprayers with dedicated fertilizer solutions are commonly used by pro growers to save time and keep organized. Some may even desire a battery-powered backpack sprayer to enhance ergonomics when working larger plots.

When it comes to managing fertility and pests (and even for micro-irrigation), the sprayer is a very powerful ally—one that needs to be used carefully. The smallest sprayer is a handheld **push-pump style** (A), which is useful for adding moisture and liquid amendments to transplant trays. **Start-up** growers may opt for slightly larger **handheld sprayers** (B) or more **heavy-duty models** (C and D) for commercial use. Having a few sprayers is a good idea; **scaling-up** growers will appreciate having sprayers dedicated to particular products (pest versus fertility products). Different colors of sprayer and labels can help you make sure you don't mix up **liquid fertilizer solutions** (1), like fish fertilizer for side-dressing melons, with **pesticides** (2), like pyrethrin for cucumber beetles. Multiple sprayers also help with bottlenecks when multiple jobs need to be done simultaneously, and they save on waste and contamination when you don't go through a

9. Garden Crop Maintenance

whole container in a single use. Most homesteaders and small-scale farmers will opt for backpack sprayers because they are so versatile. They are pretty simple to use: pour your **fertilizers** (J), like these Neptune's Harvest fish fertilizers, or scoop from **bulk containers** (I) using the **measuring cup** (3), or apply **plant probiotics** (K) like these from Terra Biosa through the **filter** (4) that is set in the **container opening** (5), adding it to the **storage tank** (6). You can then add **water** (L) to the correct ratio of dilution for the product used. If you are mixing a powder, it is beneficial to pre-mix a small amount in the measuring cup before pouring it into the container. (*Note: Don't be a fool: organic pesticides aren't harmless.* When you mix *any* fertilizer or pesticide, wear a mask, and make sure you store all products *out of the reach* of anyone except trained individuals.) The solution will move through the **pump** (7) with pump action of the **lever** (8) and pass through the tube to the **nozzle** (9). Make sure the **lid is tight** (5) and leakproof and

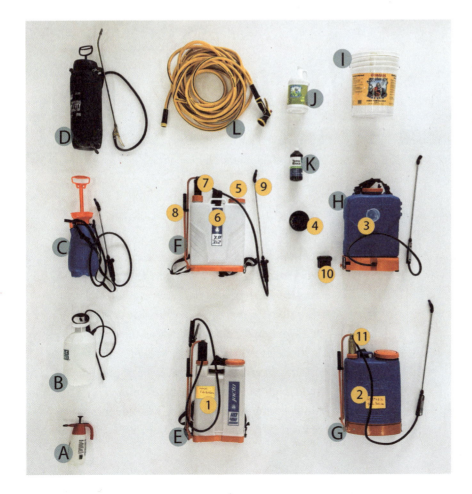

TOOL VIEW:
Fertility and Pest Management with Sprayers

then put on the backpack. ***Note:*** Test that everything is tight before you put on the backpack so there is no leakage. Don't use damaged sprayers. Some sprayers have **battery power** (10) for reducing wear and tear on the user, while others have **dual piston pumps** reaching 100 psi (11).

Jacto has a great selection of sprayers. Models like the **XP-312** (F) are comfy and lightweight, with a 3-gal holding tank that can achieve 45 psi with about eight pump actions; they are a good go-to for homesteaders. The **HD-400** (E) has a special agitator inside to help keep products like kaolin clay particles suspended in the solution, making this a good option for orchard management. This model also converts easily from left-handed to right-handed use. **Pro-up** growers may want to look at the **PJH model** (G) with a 100 psi pump for high-power liquid dispersal through the nozzle to create a wider and finer spray and corrosion-resistant metallic chamber. **Battery-powered models** (H) are another option that make long rows easy to spray without wearing out your arm from pump actions.

TOOL TIME
Suckering

Suckering can be done by hand by pinching tomato suckers, but these handy snips work nicely for larger suckers that overgrow. The **snips** even hang from your finger, out of the way, when you need your pinching fingers free again.

Pro-Tip: *Avoid using your fingers to remove larger tomato suckers because you can rip them off the main stem causing a rip down the side of the stem and exposing the plant vascular system to disease. Snips can be easily disinfected between rows and varieties. (And they should be!) Another cool tool is this thumb knife, with its hooked blade that cuts fine plant material easily while wearable and leaving you hands-free for other work.*

TOOL VIEW: Fertility and Compost

Start-up growers will use a **wheelbarrow** (A) and **compost** (B) to get their fertilizer to the garden and apply it with a **shovel** (C), cultivating it first with a **regular cultivator** (D), and using a **long-tined cultivator** (E) to properly work it in. **Scaling-up** growers will appreciate the long-term value of straw and chip mulches applied with **hay forks** (F) and **mulch forks** (G). They will also want to focus on cover cropping, especially with ryes, buckwheat, and clovers. Cover crops can be easily broadcast by hand from a **bag** (H) or with a **handheld spreader** (I). This **dump wagon** (J) from Lapp Wagons makes moving fertilizer into the garden a lot easier, but **pro-up** growers would need to make the leap to two-wheel tractors to gain more efficiency on a larger scale. Adding **mycorrhizal fungi** (K) to your garden starts, herbs, and fruit trees can help them form beneficial relationships within the soil quicker. Other **amendments** (L) and micronutrients can be mixed into potting soils and soil medium for container pots, like **chicken manure** (M), **worm casting** (N), and **kelp** (O). Side-dressing fertilizers are useful for vegetable growers, whether broadcasting by hand from the bucket, applying liquid or **soluble fertilizer** (P) from a **backpack sprayer** (Q), or using a **side-dressing attachment** (R) for **granular fertilizers** (S).

Zach hand applies fertilizer as a side-dressing to cucurbits in a new bed.

Pro-Tip: *Expensive pesticides are rarely cost-effective compared to improved cultural practices and row covers. Additionally, many pesticides (even organic ones) are broad spectrum, meaning they will harm other insects—even if applied properly. Spot application of a pesticide for a specific infestation is different from a broad-spectrum, bed-wide application.*

Stages of Fertility Management

There are several stages to fertility management. The first stage is enriching your soil as you build your garden from scratch by adding composted manures and using cover crops. The next is to top-dress your garden beds each year with well-composted compost products (weed-free), usually enriched with nitrogen or other micronutrients. The third stage is managing fertility in your transplants in the greenhouse with enriched soil mixes. The fourth is to side-dress in the field or greenhouse to fine-tune the nutrient profiles of the soil to enhance specific growth characteristics (like putting on leafy growth or flowering and fruiting). Another way to apply fertilizer is to spray it directly onto plant leaves. Foliar sprays can aid plant health and disease resistance. The last stage is *reducing* fertility inputs to avoid overgrown mature crops. Most of these types of fertility management are best achieved by improving soil from the get-go and creating a soil that helps plants get what they need when they need it. But other fertilizer applications will be needed in specific situations: when soil is less fertile in start-up years, when facing other specific challenges, or when fine-tuning your production of niche crops.

Integrated Pest Management

Integrated Pest Management (IPM) is a multi-faceted approach to managing pests. It involves providing crop protection as well as managing for soil integrity, plant health, and crop selection. **Start-up** growers' best line of defense is a good **record book** that gathers all observations about crop health during the growing season (and it should include pages for more research and solutions based on those observations). Here's an example of what goes into your record book: When you start-up, small rolls of row cover can keep pests from accessing your crops, but you may also need a handheld sprayer to make specific applications of natural pest deterrents and (if necessary) over-the-counter pesticides. You'll want to record which pesticides were used, why they were needed, when and how they were applied, and in what quantities. You'll also want to record the outcome. Believe me, you will not remember all these details come January.

Scaling-up, growers will often buy longer, wider, and different-weight rolls of row cover to better fit different situations (lighter covers for summer use on arugula to protect it without causing overheating, and heavier-weight covers for crop protection in fall against frost, for example). They may also make use of long-lasting **protection nets** (like Proteknet) and may invest in a system for **rolling up** these row covers and nets for easy removal, storage, and faster unrolling again during future reuse. Backpack sprayers are commonly used to apply products to foliage and root zones. **Pro-up** growers will increasingly use traps and insect flight schedule data to time their planting and harvest to reduce pest problems and to time the application of allowable and properly applied pesticides. For instance, sticky **delta traps** can be useful for garlic growers to see when the leek moths start to fly; **parasitic wasps** can be released as pro-grower solution. This is data that should be recorded for reference in the years to come.

Pro-Tip: *Row cover has many uses beyond season extension. It can be used for pest management, moisture retention, and frost protection. It should be set up immediately after seeding or planting for maximum benefit.*

Tunnels

Low tunnels are used to grow heat-loving crops that just need a small boost (like melons and squash), and high tunnels are used for high-value crops (like tomatoes). **Low tunnels** cost less than high tunnels. They can be made from wire or PVC or small-diameter metal hoops and covered with row cover or perforated or non-perforated poly. The terms *low tunnel* and *high tunnel* are a bit ambiguous; some consider a low tunnel to be any structure that covers a single garden bed. The grower stands in the path outside, *lifting the cover* for crop maintenance and harvest. These tunnels often run the full length of a bed (50', 100', 200', or even 300' long). Sometimes people consider low tunnels to include "**caterpillar**" and other multi-bed tunnels. These tunnels may be 16' wide (covering three beds) and require you to bend to enter from the sides; they have no end walls and only reach about 7'–9' high at the center and over a small plot of garden beds (5 to 9) (around 32'–36' wide being common widths). **High tunnels** are generally considered to be essentially greenhouses with proper framed end walls. A **greenhouse** is a high tunnel with a heating system to allow production in early spring and late fall. I have used all of these, and this is how I used them and referred to them.

My low tunnels covered one 32" Permabed Top and ran the full length of the standard bed (100' when I **started-up**, and then 300' at **pro-up**). I used galvanized steel wire hoops at 62" to cover crops like zucchini and melons for extra heat. For protection against cucumber beetles, I used a perforated poly cover, which I would remove when the plants began to spread. I used 96" wire hoops to cover crops like broccoli with either row cover or Proteknet to protect them for a longer period of time with room to grow from pests like flea beetles, swede midge, and cabbage loopers. For crops like greens, carrots, radish, turnip, etc., I just covered them with floating row cover and no hoops, so no low tunnel was used. There are also specialized wire hoops, like the TunnelFlex system, that make venting these tunnels easier for long-term production of crops like strawberries and peppers and other long-season crops. You can also make wider low tunnels using PVC pipe; these can even have rebar as support for the base, allowing the cover to go over one or two beds. These are best used with heavy-grade row cover and poly for season extension of crops like kale.

My caterpillar tunnels covered three Permabeds at 16' of total width and 9' of height at the center. They had ground anchors for tying off rope to hold the tunnel plastic down against metal hoops that are supported at their base by rebar driven into the ground every 4'–6'. I used this tunnel as a halfway point between the low tunnel and the high tunnel. It has the architecture of a low tunnel—with simple hoops supported with rope and anchors and no end walls—but it was made of metal hoops, and the interior space was tall enough for people to walk in and work, like a high tunnel. I grew the middle bed in my 300'-long caterpillar with cherry tomatoes and one side bed in peppers and basil and the other in crops like eggplants.

My proper high tunnels consisted of 20' x 50'-wide tunnels on custom-made metal skids (skis) that I could attach to a special 3-point hitch-mounted pulling bar, which allowed me to move them back and forth along sections of garden beds. I grew early greens in here in March, pulled this off them at the end of April, and started a crop of slicer tomatoes.

Then, I moved it off when tomato season was over and pulled it over already established beds of winter kale and spinach for season extension.

My greenhouse has a gothic arch that sheds snow, and it is heated for starting plants and growing early crops of tomatoes.

Supplies

Supplies should be grouped so they are easy to find, sort, and restock. This supply shelving has three primary categories: **pest management supplies** (1), **small tractor maintenance supplies** (2), and **fertilizer** (3). The shelf for pest management could include the following: **dormant oil** (A) for fruit trees, **bordo copper** (B) for blights, **fly paper** (C), **pyrethrin-based products** (D) for cucumber beetles, **BTK** (E) for caterpillars like cabbage moths, **mouse traps** (F), **water sample kit** (G) to test water quality, **lime sulfur** (H), **insecticidal soap** (I), **horticultural oil** (J), and **animal shampoos and cleaners** (K) for farm dogs. The small tractor shelf has many essentials for 2- or 4-wheel tractor maintenance (see *The Two-Wheel Tractor Handbook*). These shelves are in a climate-controlled space so can also include storage of supplies that shouldn't freeze like **paints** (L), **glues and adhesives** (M), **oils** (N), **distilled water** for batteries (O), and **other supplies** (P). The fertilizer shelf might have **liquid fertilizers** (Q), **granular fertilizers** (R), **soil probiotics** (S), **kelp meals** (T), **mycorrhizal fungi inoculants** (U), **Ziplocs for soil samples** (V), bins for small products (W), **hen manure** (X), **soil life inoculants** (Y), and **grafting and rooting supplies** (Z).

Pro-Tip:
Managing Pests and Disease
Part of the arsenal against pests—indeed, the first line of defense—should include wise crop selection for your soil and climate and best production practices. When crops are grown in the right way and suit your environment, they are much healthier and can outgrow a lot of pests, diseases, and weeds. This means sprays and other remedies can be used as spot *applications, rather than* broad *applications.*

Natural Pest Management

There are many natural pest management practices that can be implemented in your garden. Consider wild-harvesting high-in-magnesium plants, like **horsetail** (1), that can be brewed into a soil tea and used as a fungicide to help with powdery mildew and black spot. Or, grow **garlic** (2) and hot peppers to make a pest-repellent spray (together they make an unappetizing and spicy cocktail that discourages bugs and bunnies). Herbaceous thickets of some plants, like **wormwood** (3), make deterrents for cabbage loopers, codling moths, flea beetles, white flies, and carrot flies. (Wormwood, as the name implies, can also act as a dewormer for livestock like goats, sheep, and other animals.) Most livestock will eat these plants when they need them and avoid them when they don't; excess wormwood is toxic to livestock. My experience with my sheep has been that they will eat *some* and not too much, and they haven't had worms since they have had access. ***Note:*** Do your own research before applying any of these methods. My publisher and I are not accountable for your understanding and practice with these methods.

The first line of defense, however, is always healthy soil. You can improve soil health by creating your own or blending special soil teas and dry mixes of ingredients to **improve soil** (4), and you can customize fertility and structure with worm castings, micronutrients, and **soil life inoculation** (5). Remember, the soil food web—that ecosystem in the soil—is teaming with many more *beneficial* microorganisms than *detrimental* ones, and some are just waiting to help your plants find water and nutrients (like mycorrhizal fungi), shred organic matter for decomposition (like arthropods), and fix nitrogen into the soil (like nitrogen-fixing bacteria)!

Note: Another line of defense is *biological pest management*, which uses beneficial organisms to manage pests. For example, parasitoid wasps can destroy leek moths! Order them from catalogs, release them, and sit back to watch them hatch in your garden plots to manage your pest population with their own ecological balancing act of predator and prey.

Natural Pest Management

Cover Crops

Cover cropping is an important multi-stage process in the garden operation cycle. Many cover crops are seeded at the end of the garden season in order to stabilize Permabeds over the winter. This includes winter rye, which will survive the winter and grow again in spring, scavenging nutrients that would otherwise be lost to snow melt and reducing erosion of soil. Sometimes cover crops like oats, peas, and buckwheat are grown in spring and summer to smother weeds and act as green manures. Crops themselves can serve as cover crops if you just allow them to keep growing after harvest (leafy crops like lettuce and arugula work do this well) and then kill them back as green manure or *in situ* mulches.

Seed and Kill Your Cover Crops

When we talk about "turning beds over" or "re-preparing" new seedbeds (see "6: Fine Seedbed Preparation"), we are mostly talking about dealing with turning over beds that have a cover crop or crop debris and turning them back into growing beds that are ready for new seeds and transplants. So how do we get cover crops into our beds and then work them into the ground?

Summer maintenance of crops and garden beds includes **routine cover cropping** of beds, allowing your vegetable crops to actually grow into cover crops, seeding other cover crops like buckwheat and rye, and returning cover-cropped beds into production. Cover cropping is easily done by hand or with a broadcast seeder. Sometimes, seed is *undersown* around a canopy of crop like squash, or *intersown* between rows of crops like kale, so when the crops are "spent" (done producing for the year), the cover crop emerges and takes over.

Turning cover crops over needs some specific tools. **Start-up** growers may simply mow the cover crops down with their **push mower** or knock them down with a **board crimper** (think of a 2×8 attached to two upright handles) and then cover them with a **tarp** to kill them back. **Scaling-up**, growers will use **weed eaters** and **scythes** to mow their cover crop down—depending on their preference—or, on a larger scale, make use of flail mowers on the two-wheel tractor. **Pro-up** tool options include DIY crimper rollers that can be used by hand and/or using a tractor. These break the vascular system of the crops (like rye and vetch) turning them into a living mulch. For crops like salad, a great solution is a lighter-weight concrete roller that will crush and press the crop down, making it easier to smother under a tarp or weed barrier.

I use **Alternate Maturity Patterning** in all my growing, meaning crops in adjacent beds are never similar in maturity. Rather, crops like melons, which have a long DTM (80 days) are side-planted with crops that have early maturity (35–55 DTM), like lettuce, arugulas, etc. Running a concrete roller over the lettuce crop after a few weeks of cut-and-come harvesting flattens it down; then a weed barrier is pulled over that part of the bed, allowing the adjacent melons to sprawl over the clean weed barrier surface while the cover-cropped lettuce debris composts under the warm tarp, adding soluble nutrients to the soil to feed the melons.

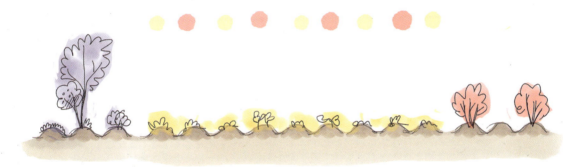

Above: *This triad (three beds together) uses Alternate Maturity Patterning, in which the squash in middle bed is 100 DTM, and the radish and arugula on either side is 30 DTM. The residue from the early-DTM crops can be left to become a cover crop before being rolled and then having weed barrier pulled over it.*

Left: *Designed for rolling over fresh concrete to help work out air bubbles and smooth the surface, the concrete roller can be used in small-scale market farming to do crimping (kill plants while leaving them in place) of light-duty crop residue. Heavier-duty cover crops (like rye) require a true roller/crimper. I have yet to see a commercially available version, but DIY versions can be made by those handy enough and inspired to use this important technique in the garden. I often use this concrete roller to help with bed packing after forming new beds to help settle the uneven surfaces and make a good seedbed.*

Grow and Mow

Mowing is important for maintaining greenspaces around your vegetable garden and when using cover crops and green manures. Cutting of plants in general can be thought of as mowing, especially as we begin to transition the way we think about "lawn" and "cover crop" and see our yards for their potential: greenspaces growing herbs, crops, and cover crops. If my lawn is a mix of thyme, chives, lemon balm, and bee balm, mowing it will release pest-deterrent oils into the air around my garden. If I harvest it, I have culinary herbs and tea plants in my harvest basket. Either way, it is a grow-and-mow management that is equally applicable in the yard as a lawn alternative, between fruit trees in an orchard, or as living groundcover between garden beds (low-growing varieties like thyme, violets, dwarf clovers, etc., lend themselves to this application). In this world, the act of mowing is also a form of harvest, as has always been the case with hay.

Start-up growers will use a **lawn mower** (A) for spaces around their garden, and, if growing on flat ground, a lawn mower can be used for smaller crop management. **Scaling-up**, growers often use handheld weed eaters like this **brush cutter** (B) from Stihl, which has a great ergonomic setup that allows hours of comfortable operation. It handles larger stems and can trim down crop debris in raised beds; it can also trim down cover crops and help clear unwanted plants—like poison parsnip—from wildlands. (***Note:*** **Wear protective gear** (C) and protective clothing when clearing wild land; poison parsnip juice, for example, "burns" your skin.) The **combi-style unit** (D) has multi heads that can be attached, including a **brush cutter** (E), a grass trimmer (which works great when working up against container beds), and **hedge trimmers** (F) for maintaining living fences. These units come in **gas** and **electric** (G)! **Pro-up** growers may shift to two-wheel tractors and make use of a flail mower to completely shred cover crops and crop debris and to manage lawns quickly.

You may wish to specialize in the art of the ***scythe*** (H). This #100R/55 cm **trimming blade** (I), at 338 g, is great for trimming soft and succulent cover crops. The #126/65 cm **grain blade** (J), at 467 g, is great for tougher cover crops and cutting grain. A **cradle** (K) is attached for grain harvest. This #012/55 cm **ditch blade** (L), at 765 g, works well in rougher conditions and is more suitable for managing raised beds. These scythes are from Scythe Works.

9. Garden Crop Maintenance   161

For container gardens, growers may opt for the **hand scythe** (M) for cover crops, the **sickle** (N) for grains, or this great **kama** (O), a rice knife made by Reforged Ironworks. The **billhook** (P) is great for heavy brush or cover crops—it is something between a scythe and a machete. For more succulent harvests, the **serrated greens knife** (Q), or these **grass clippers** (R) are suitable. **Herb scissors** (S) are what you need for snipping herbs.

TOOL VIEW:

Grow and Mow

If you are harvesting hay or any green material for curing and storage, you will need a hay fork and a rake. Larger-scale growers will find this equipment available as attachments for their two-wheel tractors.

Field Recordkeeping

Field recordkeeping is important for any grower. The items pictured are more varied than most growers might use. That's because I have a large food forest and orchard plantings as well as vegetable gardens, and I conduct research on crop hardiness and suitability for food-forest-style (or edible ecosystem) guild designs.

For annuals, you want to label the **variety** and the **date** of seeding/planting in the field to make sure employees and helpers know which crop is which. Labeling and keeping records is even more critical for those managing perennial gardens. Here is a look at the essentials I keep handy to make sure I don't lose track of the facts for my garden and my larger orchard. **Start-up** growers will make use of **plastic popsicle sticks** (A) or larger **wooden popsicle sticks** (B). I like these painted **large popsicle sticks** (C) for annuals, and I make my own **24" stakes** (D) for crops that are in the ground all year. **Scaling-up**, growers will use larger, painted stakes that don't break easily or get lost in the garden and can be reused. Make sure to write on your stake with a permanent marker *meant for outdoor use*, like an industrial **Sharpie** (E). **Paint markers** (F) are better for writing on larger wooden stakes for labeling crop and tree plantings, while **Paintstiks** (G) are great for making fast marks on stakes, trees, and rougher surfaces where a checkmark or other small mark is needed (I checkmark the wooden support stake beside trees that need disease assistance, for instance). Wide-tipped **markers** (H) are also useful for writing on larger surfaces (like temporarily writing fruit tree varieties on **upright stakes** [I]) before attaching **metal labels** (J), **ID tags** (K) and (L) with **screws** (M) using a lightweight **cordless drill** (N) or even this lightweight **USB-charged driver** (O). A **handheld driver** (P) is useful for quick tightening of metal nameplate screws when checking up on plantings in the orchard and food forest beds.

Most data gathered in the field can go into a **handheld notebook** (Q). **Pro-up** growers will often use a Toughbook-style **laptop** (R) for easy inputting in the field, or they might use their cell phones for data collection. You

9. Garden Crop Maintenance   163

will develop your own means of collecting data (laptop/notebook, etc.) and your own system of row markers and labels, but don't underestimate the power of a photo. Put photos in labeled folders on your device along with data of actual garden yields, pest issues, weed management obstacles, and any other useful information. Come January, you'll be happy to have these records as you plan your next year's garden. When reviewing my records in winter, my memory is jogged for many to-do items. For this reason, I often use a technique my dad taught: I take records in a notebook during the season on the left-hand page and then review them in winter and make my to-do lists, orders, and new designs on the right-hand page. For instance, a left-hand page might have records of tomato yields and notes about blight issues; on the right-hand page I might then make lists of winter-researched tomato varieties resistant to blight and gardening techniques to reduce blight, like growing in high tunnels.

TOOL VIEW:
Field Recordkeeping

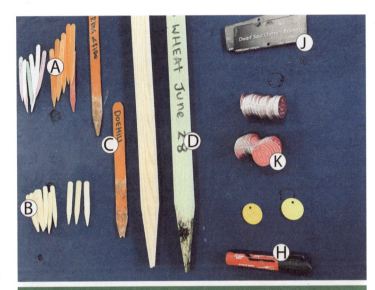

For perennials, you may want to use a **caliper** (S) to measure growth of your fruit trees or use a **soil probe** to take **soil samples** (T). There is also other work to be done while moving through the garden, like adding or securing stakes with a **mallet** (U), **pruning** (V) where needed, or using **tree ties** (W) to secure staked trees. Whatever you get up to, you need to keep your knees healthy with **pads** (X), carry your gear in your **tool belt** (Y), or cruise light with just a **mallet carrier** (Z).

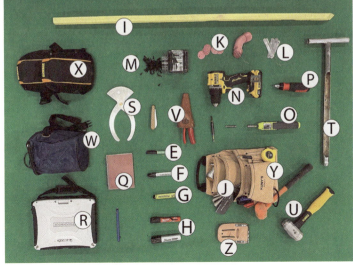

10. Crop Weeding

When it comes to managing your vegetables, there is no task more labor intensive than weeding—especially if you get it wrong! **Crop weeding** is the process of making multiple cultivation passes (weed sweeps) through your garden beds. **Weed sweep** refers to each time you pass through your entire garden and weed all rows in all the beds that need weeding. Typically, each garden bed requires **path**, **shoulder**, **between-row**, and **in-row** weeding. In-row weeding takes much more time because you have to work around young crops and not just cruise down the row with a hoe. Pro growers will lean toward **pre-weeding** and other specialized techniques like blind weeding to make this labor-intensive work less of a burden to their busy summer days. **Blind weeding** is the technique of using a tine weeder to pass over the top of the entire crop bed while the crop is growing, but the fine springy tines only harm the much smaller weeds growing in the row and between the rows of the larger, more established crop.

Row Spacing and Hoe Sizing

When it comes to weeding your crops, one of the first things to understand is that straight and equidistant rows make for fast and efficient weed sweeps. The **123 Planting Method** uses raised beds with **equidistant spacing** of five rows at 5" between rows, or three rows at 10" between rows, or two rows with 20" between, or one row occupying the center of the entire 30" bed top. Whichever type of hoe you use for between-row weeding should be equal to the row width minus 1"–3". This 1 to 3 inches is the safety zone that allows you to hoe quickly down a row without hitting plants. So, a 10" row spacing would work nicely with a 7" hoe, or a 5" row spacing with a 3.25" hoe. The straighter the rows, the narrower the safety zone can be!

The popular stirrup hoes and collinear hoes (see illustration) can be purchased with a wide selection of blade sizes that are interchangeable on one handle and collar. For years, I ran a 300-member CSA with only stirrup hoes (long-handled and wheeled versions), collinear hoes, and, of course, hilling hoes for hilling. I eventually added wire weeders to better manage weeds around delicate crops.

FOCUS
Popular Weeding Tools

Here are some of the most popular weeding styles and **hoes** (1). You can *pre-weed* and *blind weed* very effectively with **tine weeders** (A) and manage between- and in-row weeds with **spring hoes** (B) and **wire weeders** (C). **Hilling hoes** (D) are great for *hilling* and *burial* of weeds for crops like potatoes, leeks, and beans. **Furrow/hillers** like the Connecta Row Pro (E) can also do *micro-hilling*. A **tidy row** (2) of crops makes weeding a breeze. This **furrower** (F) was used to *furrow* the row to plant the onions and, turned on its side, it was later used to hill and weed them. Furrowing makes weeding easier later by providing loose soil to bury baby weeds. One of the most popular hoe designs of all time is the **collinear hoe** (3); its **narrow side** (G) allows it to slide between crops for *in-row weeding*, while its **wider width** (H) allows it to effectively make passes *between rows*.

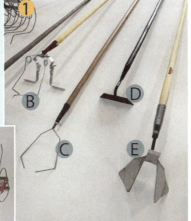

Pro-Tip: To effectively weed the entire between-row space, most growers will make two passes down a row, hugging one side of a row on the first pass and then hugging the other row on the second pass.

Between-Row Weeding

Between-row weeding is the process of cultivation in the space between crop rows. This is the easiest form of weeding, but it must be timed correctly. Crops prefer frequent cultivation to reduce competition from weeds and to loosen the soil for growth and future hilling (see "Furrowing and Hilling"), as well as to provide a dust mulching. **Dust mulching** refers to the breaking of the soil's capillary action by slicing the top ½"–1" of soil crust and breaking it up into a fine dusting of *mulched topsoil*. This has been shown to reduce water loss from evaporation and is a further benefit of between-row weeding.

Start-up growers will find hoes that are suitable to their crop type, soil, and row spacing. A longtime favorite is the **collinear hoe**, which easily slices the topsoil and allows you to adjust the hoeing width simply by angling the blade as you make your weed pass. This hoe from Reforged Ironworks with an **L-blade** (E) is a popular Slavic-designed tool used to slide underneath the canopy of crops like carrots, and it can be turned to furrow trenches for crops like beans. This **multi-purpose hoe** (F) from Grower's and Co. is a tougher collinear hoe with some meat for hilling and getting at bigger weeds. Another favorite is the **stirrup hoe** (D), with its oscillating blade that works well in soils with more stones and crop debris. **Scaling-up**, growers may wish to add more specialized hoes to their repertoire, including **loop hoes** (C) like this **omega hoe** from Terrateck for narrow row spacing (when growing five or more rows per bed). **Wire weeders** are very lightweight and won't damage drip tape or crops because they have no blade. This **wire weeder** (A) from Two Bad Cats has a soft-touch ergonomic grip and a lightweight aluminum handle. If you need to deeply aerate the soil or mix in side-dressed fertilizer while weeding between rows, I recommend a **3-tine cultivator** (G), like this one by Johnny's Selected Seeds. **Pro-up** growers will inevitably add a **wheel hoe** to their setup for its fast and efficient weeding of paths and wider-spaced rows. In my market garden, I used the **Glaser wheel hoe** (I) with a 7" stirrup hoe attachment for paths and row crops. The Terrateck **single-wheel hoe** (J) is a style I like because it has interchangeable attachments like the **cultivator** (1) for working in fertilizers. I also like the **preci-disc** (2) attachments for carrots and other small row crops. For these small row crops, I use a two-wheel hoe (not pictured) that straddles the row with its left and right wheels with the preci-disc attachment. The preci-disc has two parts: a *pair of left and right sweeps* that cut weeds on each side of a crop row, and *crop-protection flat discs* to keep the crop from being smothered by displaced soil. Another great attachment is the 4" **ridger** (3), which can furrow for single-row crops like potatoes, but I love using it to hill crops like carrots by passing between rows late in the season to prevent them from getting green shoulders (from sun exposure). *Note:* Wheeled tools like these are more suitable for extensive operations practicing row crop farming where crops are grown in flat plots with wide spacing between the crops in the row and between the rows. Wheeled tools are less suitable for intensive multi-row beds where a 32" bed top may have 3, 4, 5, or more rows per bed. Wheel hoes are popularly used in paths between these raised beds and for crops grown at 1, 2, or 3 rows in the 32" bed top, but not for more densely planted crops.

Another slick added feature on the Terrateck system is a **drip support** (not pictured) which lifts your drip tape free of the cultivating wheel hoe.

Pro-Tip: *The need for weeding passes is an important concept to grasp. You need to make multiple passes over the weeks it takes a crop to mature to keep the bed(s) free of weeds. Most short-maturity crops (~35 DTM) require 1–2 sweeps; longer maturity crops (40–75+ DTM) might require 3–6 sweeps—or more.*

TOOL VIEW: In- and Between-Row Weeding

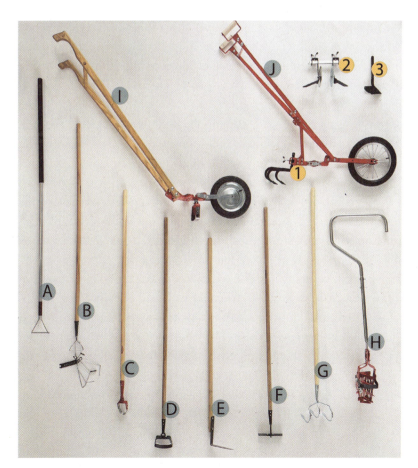

Pro growers will lean toward the increasingly popular systems for *interchangeable* hoe blades onto one handle, like the interchangeable **wire weeder** (B) from Growers and Co. that comes with different-sized wire weeder heads. When you are out doing your weed sweeps, you may find conditions have changed and you now want a different-sized hoe blade or hoe type—collinear or wire—and with these systems, the swap is just a snap and a click away! The **wheel weeder** (H), designed in Belgian Flanders, easily lifts and flicks germinating weeds early on. It is also very effective for dust mulching.

In-Row Weeding

Pro-Tip: *The best line of defense against in-row weeds is creating a proper* **stale seedbed** *(by pre-weeding) because this reduces* all *weeds, including difficult in-row weeds. This provides a* crop growth lead *(a head start for the crops) so they can germinate into a clean seedbed without weedy competition.* **What NOT to do:** *If you prepare a bed and wait a few days before seeding your crop and NEVER pre-weed at all, the weed seeds will be germinating just as your crop seeds are being seeded; this is a losing situation for in-row weeding.*

In-row weeding is by far the trickiest form of weeding. It is the process of removing any weeds that are growing *within the exact line of crop growth* and within a 1.5" strip on either side of that row. Weeding so close to crops requires precision to avoid injury to the crop itself.

Start-up growers will in-row weed by hand and use their collinear hoe to weed between crops that are planted in rows. The collinear hoe is the first hoe to buy because you can use it effectively for all types of weeding. Remember, the collinear hoe does both in- and between-row weeding effectively because its blade can be made narrower or wider just by angling the hoe. It also has a T-shape that allows you to skirt under the canopy of growing crops like carrots. Growers using a Connecta system may want to get the **Row Pro** (A), a hiller/furrower combination. If you pull it one way, it makes a furrow you can plant into; pulling it the other way fills in the furrow. Because transplants have a jump on the weeds, you can continue to manage in-row weeds with the Row Pro by hilling to *bury the weeds* in-row under a layer of soil—while leaving the crop standing above it! Growers will also use a hilling hoe, like this **asparagus hoe** (B) by SHW, found through Holden Tool Co., that can be used to hill soil into the row of a more mature crop, smothering smaller weeds.

Scaling-up, growers may add the **wire weeder** (C) that passes on either side of a row of a crop, close under its canopy of beet or carrot leaves. This delicate movement will remove small weeds and won't harm crops that have a growth lead. The **spring hoe** (D) (model shown is by Terrateck) is a U-shaped wire weeder that straddles the crop row and tickles the soil on either side; but, unlike the wire weeder, these tines overlap so it opens for the passing row of crop—*getting between the rows*. **Note:** Wires aren't sharp

10. Crop Weeding 169

and won't harm crops, but they are only effective against *young* weeds. Wire weeders and spring hoes are NOT meant for cutting deep in the soil or for hacking—like the asparagus hoe and other broad hoes.

The spring hoe shown here is great because it combines with two sharp L-sweeps that cut deeper in the adjacent space beside your crop row for *between-row weeding simultaneous to the in-row weeding*. **Pro-up** growers will make use of blind weeding techniques with **weeding rakes** (E). The one pictured here is a cost-effective model by Johnny's. It is pulled across the entire bed surface, blindly *lifting weeds out of the between- and in-row spaces* without harming established crops. Two Bad Cats makes many different sizes of masterfully designed **tine weeders** (F) from 9" up to 30" wide. Tine weeding is most effectively done by making 2–3 passes on a bed twice per week when the weeds are in the "white thread stage," meaning only the first cotyledon leaves and first little radicle root have grown (small enough to be vulnerable to a light tine tickle).

**TOOL VIEW:
In-Row Weeding**

■ Extensive Row Cropping with Wheel Hoes

Scaling-up, especially for pro growers working with a row-crop-style production (wider rows of fewer crops for market), the **double-wheel hoe** (G) will be a key tool for in-row weeding. Attachments like **finger weeders** (H) will be of interest to pro-grower row croppers (not home gardeners) and come in different hardnesses (signified by orange and yellow) for lighter or heavier soils. They work by breaking up soil in the root zone and dislodging weeds. The key is to have the crop larger than the weeds. You then straddle the crop row between the two

*Pro-Tip: For effective wheel hoe in-row weeding, it is important that the spacing between rows be 10" or greater to leave room for the wheel hoe, and you need a **double-wheel hoe** with mounting arms for a secondary tool fitting (or a **spring-loaded mounting arm** for finger weeders).*

wheels and push. The weeders turn faster than ground speed and are mounted on a spring to keep the weeder at a constant depth. This means these finger weeders are designed to efficiently dislodge weeds between the crops in the row. **Lelievre sweeps** (I) can be mounted on the double-wheel hoe for weeding between rows, or **bio-discs** (J) can be used to do some micro-hilling to cover in-row weeds, again by straddling the row. Alternatively, a **furrower** (K) can be used to pass between rows using a single-wheel hoe to displace soil from between the rows into the row of crop for hilling and smothering small weeds.

You can also combine the effect of the spring hoe and the wire weeder using the **torsion weeder** (L) attaching to the **primary tool fitting** (M), and a **tine harrow** (N) with either 5 or 7-tine harrow springs (of heavier or lighter gauge depending on sandy or stony soil) attached to the **secondary tool fitting** (O).

TOOL TIME
Wheel Hoe

The wheel hoe is a very powerful tool. It distributes the energy of the **operator** (1) standing beside or astride the crops, and all the power of your **legs, back, shoulders, and forearms** (2) through the tool framework to an **implement** that is elevated and controlled by the single or double **wheel** (3) in front. The fact that it has *implements* and *wheels* is a tell-tale sign of its advanced engineering and potential for efficiency. This tool has strong framing and wheel(s) to help *guide* and *support* the tool, and it can carry implements—it is something akin to a tractor but manual and human-powered.

A **crop row** (4) has weeds within the row itself as well as between it and adjacent rows. This is where the wheel hoe excels because it has multiple implements (like the L-blades) to cut between-row weeds while the **finger weeders** (6) lift and **flick out** (7) in-row weeds *at the same time*. To maximize the in-row and between-row weeding that a wheel hoe is capable of with a single pass, it is important to have your wheel hoe set up right. This includes having the right **finger weeder** (E) for your soil type, choosing the hoe blade for the in-row weeding that suits the situation, adjusting the finger weeder and hoe spacing correctly for the crop, and having the correct spacing for the intended weeding passes.

To adjust your wheel hoe to maximize its capability, you can adjust the **L-blade** (or other primary attachment) at **point** (A), moving it closer or farther from the crop row. The **adjustments** (B) help to release or enhance the tension on the primary toolbar, less tension provides more movement of the implement up and down when avoiding obstacles like stones in a stoney soil. Adjustments at points (C) and (D) will widen or narrow the stance of the **finger weeders** (E) or other secondary implements attached to the secondary toolbar.

Note: Wheel hoes are designed for more **extensive production**, meaning you leave more room between the rows and in-rows so the crops can be easily weeded with these implements—they succeed with great acclaim in a row crop layout where the bed system is left aside for a deeply chiseled entire plot of row-by-row planting at proper spacing.

Wheel Hoe Features

Mulch and Weed Barrier

Mulching comes in many forms. It can be in the form of natural and organic material applied to a bed surface, or it can be rolls of synthetic weed barrier. Natural mulches have certain benefits: you can make them yourself; they provide fertility and habitat for the soil and soil life when they decompose; and they don't require specialized equipment for larger-scale application. Synthetic mulches, on the other hand, offer speed of application, the ability to use hole punching and burning systems for transplanting, and very high levels of weed exclusion from cropped beds. Mulches are useful for both annual and perennial gardens.

Start-up growers will find natural mulches like **straw** (A), **wood chips** (B), and **leaves** (C) easy to work into their program. *Note:* Weed seeds can be brought in with natural material, so be aware of your sources. Hay should *not* be used as a mulch because it often has seeds.

To manage mulch, you need a **mulch fork** (D) and or **hay fork** (E). It is possible to grow your own mulch and cut it with a **scythe** (F) or a **hay rake** (G). Using a **leaf rake** (H) to gather your own leaves is worth it for growers looking to increase their natural mulch stock. *Lasagna gardening* and *sheet mulching* (when weeds in soil are suppressed by top layering with cardboard and composts) is popular for perennials. Rolls of heavy **kraft paper** (I) can **scale-up** this method. **Starting-up** growers will use narrow and shorter rolls of **lightweight weed barrier** (J). **Scaled-up** growers know heavy-duty **poly tarps** (K) are best for pre-weeding plots; heavier, wider, and longer **commercial weed barriers** (L) are great for long-term production of transplants, and **horticultural plastic rolls** (M) can be applied by hand using mulch rollers (not pictured) or two-wheel tractor-mounted mulch rollers.

Popular mulching material for perennials includes coconut husk mulches and other natural fiber mulches that will slide around perennial stems. Some of these come as squares or rounds of mulch that slide around a tree stem via a small slit from the edge to the center, like these **hemp mulch squares** (O). **Ground staples** (P) help to hold rolls of plastic mulch down; row bags (not pictured) also work. Watch the wind. Mulch can be less effective (or useless!) in high-wind areas if not applied and weighed down properly. In high-wind areas, an *in situ* mulch is beneficial; you grow a rye and vetch cover crop and roll/crimp it down as a mulch. ***Note:*** Synthetic

10. Crop Weeding   173

TOOL VIEW: Weed Barrier and Mulch

Bottom left: Zipperbeds use synthetic mulch to supress weeds. No holes need to be cut in it, so the plastic roll can be used again. This type of mulch can be removed once annuals or perennials are established.

Bottom right: Organic mulch like this second-cut hay mulch provides nitrogen and carbon for the soil and decomposes to create a fertile soil for the next crop.

Synthetic Mulch

Organic Mulch

mulches can be purchased with pre-made holes, or you can burn the holes in them with whatever spacing you like. *Horticultural plastic mulch* is designed for use with dibblers to pierce it, but it can also be hand cut. Any of these rolled mulches can be applied to the Zipperbed method: two pieces are laid, one on either side of a row of perennials or an annual crop. No alteration of the material is needed, so the rolls can be reused, which saves on costs and reduces pollution.

Carts, Wagons, and More

Getting around a property with your tools, supplies, seeds, crops, etc., requires *transport*. **Start-up** growers will rely on a wheelbarrow or **dump cart** (B), like this Smart Cart. **Scaling-up**, growers will often get a **Vermont-style cart** (A) which accommodates harvest crates easily and has a removable back for easy loading and unloading. **Deck wagons** (C) are also popular for moving transplants and other essentials; there are multi-deck versions for pro growers. **Heavy-duty dollies** (D) are useful for moving supplies and helping to load market trucks. Lightweight and **foldable dollies** (E) are great to carry with you to events and farmer's markets. **Pro-up** growers will often want a **pallet jack** (F) for moving pallets and bulk crates. On larger farms, these can be moved by small tractors, replacing the pallet jack with pallet forks as an attachment on their tractor loader.

Pro-Tip: Custom designs that fit your standard supplies can really maximize a cart trip or shelf storage space. You'll never be able to buy anything better than something you build yourself to fit your standard bins, crates, and other containers! (But, it's best to be at your static scale before you invest in the DIY build.)

10. Crop Weeding

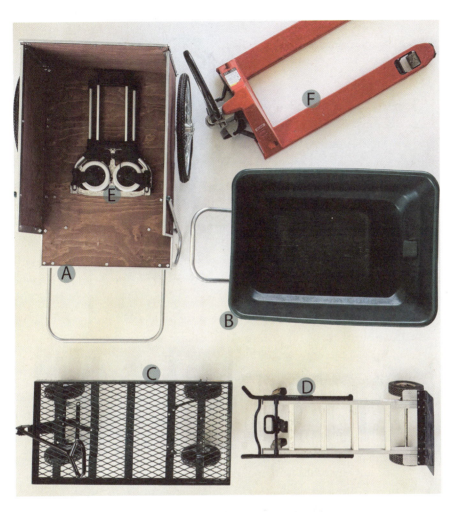

Mesh-Bottom Carts

Bottom: *These Vermont-style carts at Beaverland Farms were built with mesh floors to allow water and debris to fall through. This is a great design because it is the floor of a wooden cart that usually rots first. These carts are a custom size that fits Paperpot trays nicely, and they haul more bins than smaller carts would (up to 4–6 bins in one go).*

11. Garden Harvest

Harvest is (obviously) a key stage. It may be the one where scaling-up can most greatly enhance efficiency and crop quality. When growers increase their production as commercial market gardens, the weak link that usually develops is in the harvest production stage because it is a complicated and time-sensitive set of tasks. Crops must be harvested within certain windows of time to maintain quality, and the correct timeline has to be followed to get produce to market while it is still fresh. As such, any tools that improve crop readiness for harvest and speed up the harvest itself are a blessing. Homesteaders who grow many bulk crops for cellaring will also find they benefit from more professional tactics in their fall harvest tasks.

Start-up growers can make use of **digging forks** (A) for carrot and root harvests and **knives** (B) for greens. At this stage, practical **rubber totes** (C) are good for holding crops, and knees are kept happy with **knee pads** (D). **Scaling-up**, growers will still use many of these tools, but they may specialize with more specific harvest tools: **billhooks** (E) or **sickles** (F) for heavier perennial greens and small grains or herbs, and **garden/herb scissors** (G) to cut basil and other delicate greens. **Berry pickers** (H) are specialized tools that may apply to some crops. **Lettuce knives** (I) suit salad growers, and curved and **serrated knives** (J) are great to help harvest round, stalky plants/greens. Some models of these tools have non-slip grips, which is a nice feature. A **Salanova cutter** (K) from Johnny's Selected Seeds is a very specialized tool developed for their Salanova greens—which is great if you're growing those greens. **Scaling-up** often means ensuring you can easily pick up your crop; for this, a **sturdy cart** (L), **crop-specific crates** (M) and **tubs** (N), and **wearable picking bags** (O) are crucial.

All growers should keep good harvest **records** (P) based on their initial crop plan. **Pro growers** will add efficiency, which usually means mechanizing very specific aspects of harvest for the crops they specialize in, like the **quick-cut greens harvester** (Q) for salad production. A **cordless drill** (R) powers the quick-cut harvester, and you shouldn't forget to keep the moving parts **lubricated** (S). Pro growers will often use a two-wheel tractor

11. Garden Harvest   177

and cart to speed up transport of bulk crops over the longer distances that come with increased bed space.

TOOL VIEW: Harvest

FOCUS

Crates

Crates are essential for harvest. They provide a means of getting crop out of the field, allow water and air to flow around harvested crops, and are easy to clean. A great harvest crate will lock into a stack in the cart and may even have a "nesting" feature, meaning they sit inside each other when empty to take up less space in storage. These yellow crates at Cully Neighborhood Farm in Portland, Oregon, stack and nest. These black harvest crates with tomatoes don't, but they were purchased as older models for only $2–$3 each. **Start-up** growers will look for deals; **pro-up growers** will put more value on storage space savings because they need hundreds of crates. Nesting crates and collapsible crates (not pictured) can be broken down as you make your market sales, and they end up in a tidy stack at a fraction of the original volume. Smart growers will look for *used* stackable/nestable crates.

TOOL VIEW: Crates and Bins

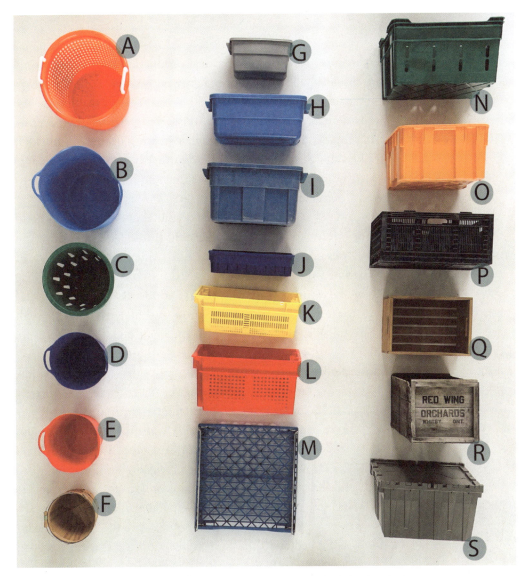

A. vented harvest container (1.25 bu); B. Gorilla tub (10 gal); C. vented bushel picking basket; D. Gorilla tub (3.5 gal); E. red Gorilla tub (3.5 gal); F. half peck orchard basket; G. salvaged mini tote; H. Rubbermaid tote (38 L); I. Rubbermaid tote (68 L); J. berry picking crate; K. harvest lug (23.6" x 15.7" x 7.8"); L. harvest lug (23.6" x 15.7" x 11.8"); M. repurposed bagel tray; N. stack/nest harvest container (1.75 bu); O. utility harvest container (1.22 bu); P. collapsible harvest crate; Q. new vintage-style orchard crate; R. salvaged original orchard crate; S. attached lid container.

12. Season Extension

Season extension is one of those critical production stages for growers in colder climates, allowing a longer growing season. For some crops like cold-tolerant spinach, kale, and hardy lettuces, this can increase the weeks of available harvest for home use and commercial sale by covering and protecting these crops against the harsher weather in the autumn. For other crops like watermelons, blue corn, and other heat-loving crops, it is a means of protecting against late frost in spring and increasing heat retention at night for better spring growth.

Start-up growers may choose to build a cold frame out of wood and old windows or opt for something like this **cold frame** (A) by the Austrian company Juwel, designed to help vent heat easily when needed. In place of the square boxes of wood and aluminum used for cold frames, it is common to employ **galvanized wire** (B) or **PVC conduit** (C) or large **metal hoops** (D) to make low tunnels when **scaling-up** and using larger high tunnels or caterpillar tunnels as you **pro-up**. Wire and PVC hoops are usually combined with **row bags** (E) or **row cover hand pegs** (F) to secure either **perforated poly** (G) (recommended for crops like zucchini, melons, and other heat-loving crops) and **row cover** (H). (Lightweight row cover can "float" atop crops like arugula and radish to protect them against pests and provide warmth, but heavy grades need to be used over hoops for frost protection.) **Weed barrier** (I) is also used for season extension because it increases the heat in the soil, allowing crops to grow faster earlier and stay productive longer. When managing heat, especially in high tunnels and greenhouses, a **shade cloth** (K) may be applied to reduce light transfer during the warm season (it is removed later in year). **Scaling-up**, growers will buy these products in full rolls of over 300 feet; smaller rolls may be available, but they are usually of a lesser quality. Consider buying from commercial suppliers. You can cut what you need and distribute the rest among other small-scale growers and backyard gardeners. Plastic sheeting and

Low tunnels made of galvanized wire hoops and row cover help protect watermelons from late frost and increase warmth retention at night for better growth in springtime.

180

12. Season Extension

row cover can be attached to the hoops with **C-clamps** (L) to PVC conduit hoops or with **cords** and **ropes** (M) that serve as the purlins and horizontal support in caterpillar-style tunnels. This allows fast venting of tunnels when combined with **J-hooks** (N) that can hold the poly up periodically along the length of a tunnel. Caterpillar tunnels also need **ground anchors** (O) to tie off the horizontal **tie-down ropes** (M), and **rebar** (P) to secure the metal hoop in place to the ground. Caterpillar tunnels have the advantage of being very low cost per interior square footage of warm tunnel space because they are secured into the soil only with ground anchors, and the hoops are supported only by rope purlins and simple tie-down ropes to secure the poly cover onto the hoops. But this low-cost assembly also makes them vulnerable to high-velocity winds. **Pro-up growers** (especially

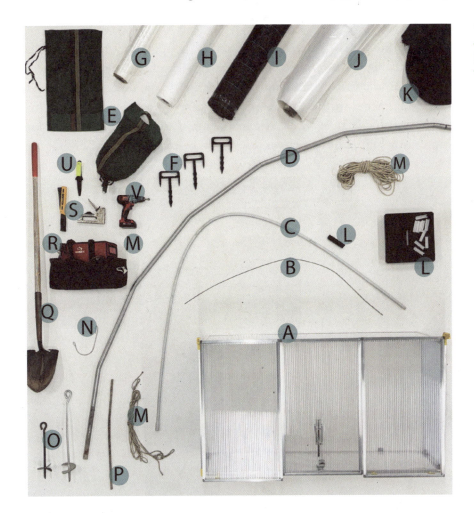

TOOL VIEW:

Season Extension

in windy locations) will opt for sturdy metal high tunnels with framed end walls. They will also employ row cover wind-up systems (not pictured) to make applying and rolling up field covers for storage much more efficient.

Other essential tools in your season extension kit include a **long-handled shovel** (Q) for applying soil to seal row cover against cold and pests (used separately or in conjunction with row bags). In your **kit bag** (R) you should always have **good staplers** (S) for stapling poly to cold frames, a **fisherman serrated knife** (U) for cutting covers, and a socket wrench and/ or **cordless drill** (V) with bits for screwing and building frames and the correct sockets for your nuts and bolts, and hex head self-piercing screws for connecting purlins and other tunnel construction and repair.

DIY Hoops

You can make your own wire hoops (and PVC hoops, too). Order bulk rolls of .135" diameter galvanized wire and cut into lengths of ~60" to 100", depending on the crops you intend to protect. PVC conduit normally used for running wires with .25 to .5" diameter can be used to make hoops for larger low tunnels than wire can achieve (and these are often preferred for winter growing). A **hoop bender** can be purchased, or you can DIY one for

Right: Small hoops can be bent from galvanized wire, pvc pipe conduit, and metal piping.

making hoops of different widths and heights. (Widths of 3', 4', and 6' are common, but benders for larger high tunnels can also be made.) This DIY hoop bender (above) at Cully Neighborhood Farm in Portland, Oregon, is a great example of a simple and effective design for affordable hoop house construction. Integrated into the wash station framing, this tool is always where it needs to be when hoop-bending time comes around!

Above: *A caterpillar tunnel in PEI protects salad greens over the winter with smaller hoops and floating row cover. The high tunnels at Beaverland Farms in Detroit extend the season for heat-loving crops like cucumbers, tomatoes, and peppers.*

Left: *Floating row cover in my market garden fields can be used with or without hoops, the lighter the row cover, the easier it floats over a crop to protect it from pests, winds, and desiccation. It keeps out the cold and holds moisture and warmth for germination and crop growth.*

13. Post-harvest Handling and Curing

Once the crop leaves the field it enters a new production stage called *post-harvest handling*. This stage can include many different tasks, like washing and drying greens, spraying off roots, and cleaning and tidying crop for market. Some of the tasks are so large and span so many weeks they are really their own type of post-harvest handling, such as the curing of crops like garlic.

Crop will come in from the field in crates, **picking bags** (A), **vented tubs** (B), or other containers. **Start-up** growers will find they do a lot of hand spraying with **spray guns** (F), so they would benefit from good **wash gloves** (E) and **waterproof work wear** like a jacket (D) and **overalls** (C). I have been using the ones shown here (made by Viking) for almost 20 years. I like employing **bagel trays** (G) to hold my **rubber band** or **twist-tied** (H) bunched carrots, beets, and radishes because I can plop them down on top of my slatted wash table and rinse them off without having to move them from the tray. Stacked bagel trays also serve nicely as curing towers for crops like garlic, onions, and squash. I do curing upstairs in my barn where **high velocity fans** (I) blow easily through the well-spaced crop to disperse moisture and help drying-down.

Scaling-up, growers will acquire (and/or make) more specialized tools, like this homemade **garlic bulb sizing board** (J) or the thorn stripper for flower farmers seen here, above the **pruners** (K). Other garlic post-harvest handling processes include trimming roots with scissors and cutting the stalk with a pruner (K), weighing the harvest on a small **kitchen scale** (L) or even a larger shipping scale, and then bagging the garlic for sale in small **mesh socks** (M) and large **wholesale bags** (N), both of which are suitable for longer-term curing, as they let air flow through them. They can be kept in these until sale or used for seed. **Pro-up** garlic growers will have their bags sized for sale dialed in and make use of quality labels that can be stapled to these small mesh socks. They will also employ a variety of **box sizes** (O) that can be sealed by **tape gun** (P) for shipping to customers. Organization is paramount, and **painter's tape** (Q) can be folded over drawstrings on the large **mesh bags** (N) with the variety names and date in Sharpie to prevent variety

13. Post-harvest Handling and Curing 185

mix-ups. Health is something we prioritize far too late. Although pro growers are usually the ones who start to worry about this, **start-up** growers handling dusty crops like garlic should buy quality **dust masks** (R) from the get-go.

When managing wet crops, like salad greens, the **start-up** grower often spins them dry with a **dynamic salad spinner** (S), either the **manual** (T) or **electric** (U) model (electric is more popular). **Scaling-up**, I used the manual model before retrofitting an old washing machine to spin my lettuce dry! I put clean salad greens loose in special "clean" **Rubbermaids** (V) before bagging them in **bags sized for ¼ lb of loose greens** (W); other crops go into the **potato bags**, shown next to these. Finished crop can go into handy bins for market, like these **crates with attached lids** (X). Some crops skip the entire post-harvest handling process because they are harvested directly from the field using a berry picker into **pint boxes** (Y) already lined up in a **berry harvest crate** (Z), which is moved directly into a dehumidifying fridge or cold storage before being processed, enjoyed at home, or sold at the market.

TOOL VIEW:

Post-harvest Handling

FARM FEATURE

Post-harvest Handling at Fisheye Farms

Andy Chae and Amy Eckert have been running Fisheye Farms in Detroit, Michigan, for the last 9 years. They produce enough on their 2 acres to sell CSA shares, as well as at markets and directly on the farm. Their **wash/pack facility** (1) is made of shipping containers framed with a generous roof. The 40' cube shipping container is used as their "clean" storage space for tools and supplies for market set up and events. They built the roof to cover the whole space and used polycarbonate windows to let light in and keep weather out. The **20' shipping container** (C) has been turned into a walk-in cold storage. They used spray foam between the original steel studs and extruded polystyrene panels as an additional solid insulation sheeting to leave no thermal bridges. They are able to maintain this space at 40–50°F, which is ideal for crop storage. In the **field office** (2), they use a salvaged **whiteboard** (E) for planning and team communication. Items on the board include weekly orders harvest lists, *To Do* lists, purchase lists, employee notes, common tare weights for packing orders, and overall inventory notes. **Note:** Whiteboards are commonly used on farms to keep everything on track. Some information is routinely added and erased, while parts of the board hold permanent, seasonal information, like the crops grown and names of markets, restaurants, and clients that make orders. The **salad workstation** (3) includes this DIY **salad drying table** (G) made from light gauge (red) steel tubing and a mesh frame made of recycled plastic 2x4s that won't rot. The whole mesh top rotates to easily get rid of any lettuce debris. **Note:** This is an important feature to avoid little bits of dried salad residue from getting caught up in future fresh loads of clean salad.

There is a 24" box fan (not pictured) cantilevered over the table to assist in drying the greens. The **root washing area** (4) includes this 12'-long **root spray table** (I) made of 2x4 framing with ¼" hardware cloth. This does double duty as a spray table for root vegetables and as a cleaning station for washing and drying their copious rubber totes. The **salad washing station** (5) also has a *salad bubbler* made from a Growing Farmers kit and plans that include a heavy-duty (new) 100-gal poly **livestock trough** (K) equipped with a **Jacuzzi blower** (L) that sends air into a **PVC manifold** with drilled holes (M) to aerate the water and move it around the cut salad being washed in the tub. An additional **bottom drain** (N) allows quick drainage for cleaning between salad wash days. The salad *spinner* (made from a Growing Farmers kit) was made from a (new) **top-loading washer** (O) with the center agitator component removed and a 20-gal **food-grade garbage can** (P) put in its place to hold the salad for spinning. Holes were drilled for drainage. **Andy** (7) fills the wash tub by hand, but as he continues to **pro-up** his equipment and fine-tune his stellar wash area, he plans on automating the filling with an auto on/off fill valve because a farmer is always on the go on harvest

13. Post-harvest Handling and Curing   187

days, and the temptation to drop the hose in and leave it to fill can result in over-flowing! Other tools and equipment they want to add in the future include a refrigerated box truck, improved shelving for the containers to make storage more efficient, a computer and printer in the barn, a floor drain in the wash/pack area, and an enclosure for the entire space—for warmer winter washing!

The wash station setup at Fisheye Farms in Detroit includes homemade and purchased tables for washing and packing and a DIY system for washing greens in a new cattle trough and spinning them in a redesigned washing machine. This is a normal scene in small-scale market gardens, as post-harvest handling is one of the important phases in which saving time with tool innovation and acquisition is key.

FOCUS

Pro-Up Is Often about Post-harvest for Commercial Growers

Pro-up grower Chris Jagger is also a larger-scale producer and has a post-harvest handling system that handles large volumes of greens, roots, and fruits for large CSAs and farmer's markets. The **high tunnel hoop houses** (1) grow a variety of **greens** (2) in the spring using the Jang seeder with the X12 roller and a 9/13 gear ratio (see "How to Calibrate Your Jang Seeder"). While the weather outside is still limiting quality production of salad crops, the high tunnel is booming! Chris seeds greens in rows by variety instead of all rows in one variety per bed, making seeding faster and harvest of a single row more efficient. This also allows him to avoid harvesting rows of varieties that may have overmatured and focus on varieties that are of the best quality. A **magnetic bar** (3) is a great tool for organizing metal tools, especially all the blade types he needs to suit different crops. The **curved blade** (A) is used for broccoli, the **trapezoidal blade** (B) is an *end push-style* blade used for head lettuces, the **rectangular blade** (C) is used for cutting salad greens and other mixed greens, and the **little knife blade** (D) is used for zucchini (it's able to fit in between the big succulent leaves without damaging them). Specializing your knives when you grow a lot of a crop makes a world of difference in speed and quality.

The **greens wash line** (4) can have bunched greens fed through it, but for wholesale, Chris ran entire boxes through and flooded the whole case. This model has an expandable and moveable roller table (D) that can stretch to hold 10–12 cases of greens, allowing for easy top icing after washing.

The **brusher washer** (5) with a **circular sorting table** (E) is used for beets, turnip, celeriac, daikon radish, potatoes, and even cucumbers.

The **salad greens wash basin** (6) has a **DIY manifold** (F) with a **hot tub blower** (G) for agitation of the greens with air bubbles generated inside the basin from the inserted PVC piping. This helps improve the thoroughness of the washing, removing dirt and debris without bruising the salad greens.

Packed boxes of crops from market await delivery (7). In the background, the **ice machine** (H) generates the ice for top icing produce to super cool it for delivery. You can also see the **water storage tank** (I) used for supplying wash water. This tank is kept in the barn to keep water at a cooler temperature; on hot days ice is added to the wash water.

These **DIY drying racks** (8) have been cleaned and are drying in the sun. Chris uses these racks in an old walk-in cooler equipped with a large DIY dehydrator to dry bulk peppers for sale.

Chris (9) reflects on years of building up his post-harvest system. He rests against the Scott Viner carrot root harvester, "an engineering feat of genius," that helps his harvest workflow keep up with his post-harvest processing capacity.

13. Post-harvest Handling and Curing   189

Check out Chris's podcast: Not Only Farmers.

One of the smallest supplies is the lowly rubber band, but it is a vital tool on harvest day. The trick is sizing the rubber band for the type of crop so just the right minimum number of wraps is needed to maintain bunch integrity. #8 is a popular size.

190 *The Garden Tool Handboook*

FARM FEATURE
Curing at Orto Vulcanico

These prized tomatoes are grown on the slopes of Mount Vesuvio in Italy. They are grown in volcanic soil and stored in **wooden boxes** (1) and are diligently tested for quality as they **cure** (2) by master gardener father/daughter dynamic duo at **Orto Vulcanico** (3). The cured tomatoes are **braided** (4) and sold hanging in wooden **display boxes** (5), loose in **baskets** (6), turned into prized **sauces** (7), sold by the **liter** (8), and packed hanging in a **custom box** (9) for shipping across Europe. These various and beautiful packagings show that tools and supplies can be specialized for your market needs.

14. Cold Storage

Root Cellaring

Traditional cold storage, called *root cellaring*, makes use of the natural temperature modification of the earth. **Start-up** growers can experiment with low-cost methods, like this small **dugout root cellar** (A) into the side of a hill. It was constructed of just cedar logs, earth, a waterproof membrane, and a metal screen. In winter, heavy-duty bags full of fallen leaves block the doorway. **Scaling-up**, growers may find the popular CoolBot system quite useful (see "CoolBot System," below). I have built several of these systems; in fact, it was my first cold storage system after using simply spare fridges. **Pro-up** growers may be satisfied with this setup depending on the scale of their operations. When they need larger spaces, growers look toward having larger walk-in cellars that can handle pallet jacks and other large equipment.

In 2007, I designed and built a **passive cellar/cold storage** (B) that had zero electric needs other than solar-powered lights. Its features include salvaged **prefabricated concrete arches** (4), multiple **temperature/humidity bays** (5), an **airlock** (6), **perimeter drainage** (7) that stores run-off in a pond, and an **ice chamber** (8) that provides 365 days of cooling using state transfer of ice to liquid. The ice-cold thaw water is **collected** (9) and held for further cooling before

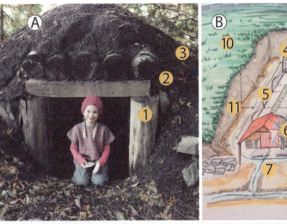

TOOL VIEW: Root Cellar

A root cellar can be as simple as a hole in the ground (A) or as extensive as the large-scale in-ground cold storage (B) I built to house my harvests year round.

being drained into a pond system for future irrigation of crops in the field. The whole cellar is dug 45' into the **glacial moraine hillside** (10), and the floor is **16' below ground** (11). An outdoor shade keeps the heat off the face of the cellar.

I won an Agri-Innovation award for this cellar; it is a hallmark of the sustainability humans used to rely on for cold storage—with a modern twist. I never cut ice from a lake; I have a giant ice cube tray I built using **Rubbermaid bins** in my wash station (12) nearby to make the **ice** (13), and I use an irrigation system to fill the cubes for easy workflow. The cellar maintains 32–39°F (0–4°C) all year round and allows storage of **root cellar crops** (14) in winter and fresh market garden veggies in summer. I have used it for an event space for **in-cellar shopping** (15) with great success. I also use it for tree sapling storage for my edible ecosystem installations— along with all my root crops. Since the ice (13) lasts from the time it's made in January till beyond when it is even needed (in October), the cellar will maintain ideal conditions of 32–39°F (0–4°C) and 90% relative humidity, making it great for storing any summer crops and winter storage crops that like the cellar conditions.

CoolBot System

Most market growers **scaling-up** will build a small walk-in cooler that is **heavily insulated** (A) and makes use of a **CoolBot**, an innovative device that commands an **over-the-counter AC unit** (C) into performing commercial cooling. The cold storage itself is best designed as a **linear space** (D) to allow for easy movement and storage of **bins of produce** (E) on **racks** (F) and in **coolers** (G) and other **stacking containers** (H). Cold storage should have **vapour-proof and shatter-proof lighting** (I)—you don't want glass breaking and falling into an open produce bin, ruining the lot. The CoolBot's **main hub** (1) is a **controller** with **temperature readings** (2), **settings** (3), and connections to all its main components: **room temperature sensor** (4), **fin sensor** (5) (that helps prevent the AC from having condensation freeze-up), **heater cord** (6), and **power supply** (7). The heater cord (6) gives a false room reading to the **AC's room temperature sensor** (8) by providing heat to it; the two should be wrapped together with tinfoil for best results. Usually, an AC won't cool below 60°F (16°C), but with

14. Cold Storage   193

the right brand of AC, proper insulation of the cold room, and the correct BTU sizing for your space and a CoolBot, the temperature can be brought down to 34–40°F (1–4°C) for different vegetables, cut flowers, etc. *Note:* These cold storages don't create humidity, so, unlike in a root cellar, produce is stored in closed bins so it doesn't dry out.

TOOL VIEW: Cold Storage

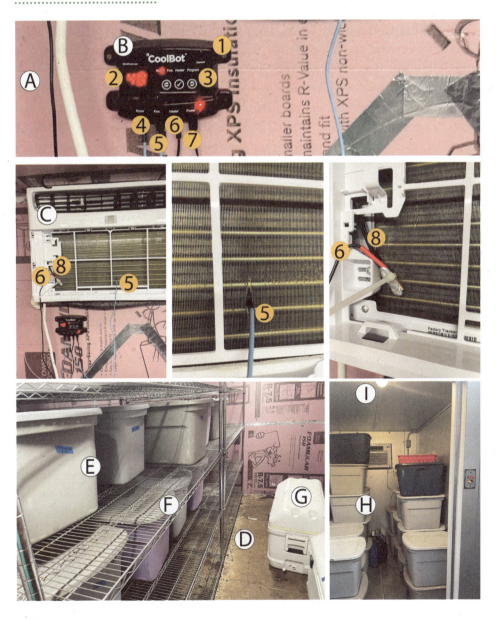

15. Marketing

Not all growers will go to market. For some, growing is a home production for enjoyment, for others, the focus may be on Community Supported Agriculture programs (CSA), or online sales. In all cases, the benefits of marketing your crops (including produce processed into shelf-stable products) include opportunities for sales, barter, and community connection. A lot of time and effort goes into marketing. Here is just a glimpse into the world of tools for marketing your crop.

Your crop will need to travel to market. For that, sturdy Rubbermaid bins, coolers, and crates (see "Tool View: Crates and Bins") are used to fill up a car, pickup truck, or van. Once you arrive at market, up goes the tent or you find your spot within the supplied booths. **Start-up** growers will make use of **foldable tables** (A) with a **tablecloth** (B), maybe with a **country look** (C). (Covering the table first with **burlap** (D) that goes almost to the ground keeps the over-curious from peering into the bins stored beneath the table.) Small **chalkboards** (E) and **larger boards** (F) can help market your products, while **pints and quarts** (G), **peck baskets** (G), and **wooden crates** (G) can help display product. **Scaling-up**, growers may bring plywood to use as an extra deep table, supported with **foldable sawhorses** (H). Tiered display shelves or **antique ladders** (I), help create a layered display, which tends to attract customers. Custom info signs, business cards in holders, and **custom labels** (J) make your products and farm memorable. And don't forget to design a nice **banner** (K) with the farm name. Styles that attach with Velcro or clips to a tent or table sides are popular. **Pro-up** growers will often make custom displays, such as this specialized **maple bin** (L) with a liftable lid that supports laminated **informational signage** (M) while holding sellable items (like heirloom garlic). The benefit of these bins that have exact measurements to fit into transport vehicles and on market tables reflects the nature of a multi-functional tool that eliminates stocking the tabletop (just lift the lid and sell). Growers can also display **award plaques** (N) and make use of custom backdrops like this **Mongolian yurt lattice** (O) that travels easily in a vehicle and expands to connect to tent stakes or booth walls using various **bungees**, **zip ties**, and **rope** (P) and

15. Marketing

can even be used for hanging braids of garlic using S hooks (or bent wire). **At any scale**, marketers will appreciate a **simple seat** (Q) and **marketing attire** (R) that is comfy, warm, branded, and utilitarian, as well as mugs for coffee, insulated containers for food, and drinking **water** (S). Marketers will need a **cash box** (T) and/or money apron. Other key supplies include **biodegradable/reusable bags** (U) for customers, a sprayer to keep bunched vegetables fresh, and twist ties and rubber bands for fixing broken bunches. Tape and a tape gun can help when **boxing** (V) or bagging orders. Pro growers will want a **professional scale** (W), and they should keep **solid records** (X) for all market sales, with notes about weather and crop quality. A kit of other essential tools might include **scissors**, **pruners**, and a **knife** (Y) as well as a **flashlight and headlamps** (Z)—because most growers get up before the sun to go to market!

TOOL VIEW:

Farmer's Market

Every market garden (1) will focus on crops that are great for market, like salad greens (2). Some can be displayed as is (3), while other crops benefit from packaging for display and longevity (4). Processed goods (5) are also popular and can be great treats at an on-farm event or farmer's market (6). Having farm-to-table food (7) available at an on-farm market is a big draw; visitors can shop for fresh produce (8) while also enjoying community interactions (9).

15. Marketing

A good market setup will have a variety of containers and types of packaging to help draw customers in and showcase your products. Shopping should be an efficient experience. Make it easy for customers to find what they want and maximize your limited booth space and time window to bring in weekly income.

16. Cleanup and Maintenance

There is a lot to clean up at the end of the season—even if you've done your best to clean as you go along. "Clean up" means numerous things: cleaning up debris in the fields; making compost; cover cropping; cleaning up workspaces, like wash areas, pack houses, and workshops; and the cleanup and maintenance of your tools, equipment, and supplies. Each of these has its own set of tools and routines to consider.

Field Cleanup

Field cleanup includes removing crop residue and creating compost. This is one of the most important seasonal tasks because it deals with an obstacle for future production (debris) and makes a key resource (fertility and soil health). **Start-up** growers may focus on forking for a lot of **cleanup work** (1). They may remove crop residue with a **garden fork** (A), moving it into wheelbarrows or carts to haul it to compost piles or composter bins. They may also want to employ **mulch forks** (B), or **hay forks** (C) to remove debris or mulch beds over the winter with straw to protect and build soil life. **Scaling-up**, growers may fine-tune their fall bed preparation by employing **broadforks** (D) or **digging forks** (E) to subsoil their beds before cover cropping them over the winter. Sometimes, specialized forks like the **rose fork** (F) or other small **hand forks** (G) will work better, especially in tight perennial beds in the food forest. DIY **compost pile systems** (2) are popular for home-scale growers and community gardeners, but a purchased system can also be useful for **smaller-scale producers** (3). As you scale-up, try to build systems for better turning of your compost, keeping it aerated, and maintaining the right temperature (use a thermometer!). You might even add drip irrigation to keep the pile moist for improved decomposition. **Pro growers** will have systems to maximize easy garden cleanup and composting of debris. One option is using large weed barrier to cover crop debris and cover crops to smother and fry them into *in situ* **composts and mulches** (4). On larger-scale farms and homesteads, field cleanup will entail using equipment like a 4-wheel tractor to move debris out of the field

16. Cleanup and Maintenance   199

and turn compost piles. A two-wheel tractor might be what a market gardener needs to flail mow debris right on the bed top and turn the bed over.

FOCUS

Compost in Paths for Debris and Fertility Management

On my farm, I use a number of different methods to clean up crop debris and make compost. In my medium-scale gardens and food forest, I have high debris output. I use the **Compost-a-path system** to make *in situ* compost out of the debris (meaning I don't have to move it, pile it, turn it, etc.). In this system, debris is pulled into wide paths and flail mowed into the soil, and the land is then cover-cropped. By not moving debris around (and with it, any possible disease), I save my back, and I end up with compost just where it is eventually needed as fertilizer. Every 2–3 years, the path material is turned back onto the bed top as compost.

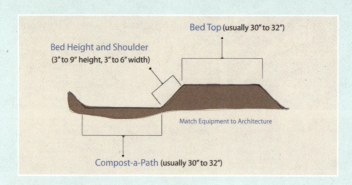

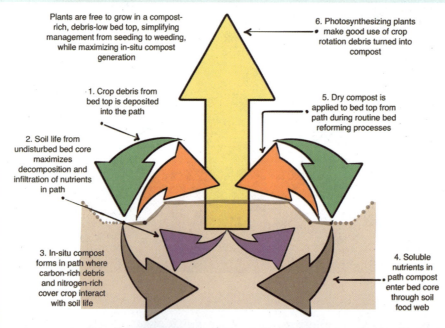

TOOL VIEW: The Permaculture Farm

There is a lot of garden cleanup on a Permaculture farm or for edible ecosystem-style gardens and food forests. Why? Because there is a lot of debris to turn into compost. What does it mean to be a Permaculture garden? Diversity?

Vera is a serious suburban homesteader. She does most of her own starts in her sunroom (1) using smaller 1010 trays with humidity domes to **germinate** (A) and a variety of **cell packs** (B) and pots for final transplants after potting up. **Wire racks** (C) are multi-functional and can be set up for doing her starts and dissembled to free up space for the rest of the year. She grows a lot of tomatoes, corn, melons, pumpkins, garlic, and greens. This is a family affair (2), and Grandma, who grew up on a rural property, is a regular farmhand at Vera's suburban yard. Her son Alex is also an avid gardener and has ventured into selling his own produce at a roadside stand.

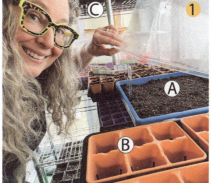

Vera is a Permaculture homesteader; she uses edible ecosystem design to achieve bounties of both **annuals** (3) and **perennials** (4), and she forages in her yard and adjacent wilderness for other bounty, such as **edible fungi** (5), including favorites: **boletus** (C), **chanterelle** (D), and **lobster mushroom** (E).

TOOL VIEW: The Permaculture Tools

The Permaculture farm or homestead has many tasks, tools, and techniques that are different from those on a more typical garden or small farm. With the focus on more than one enterprise and usually integrating very different lifeforms in their productions (perennials, fungi, animals, and gardens), these growers acquire tools that meet their diverse needs. Take, for instance, a food forest grower who has a much greater need for tools more typical to forestry, beekeeping, and small livestock alongside their garden tools. As a Permaculture homesteader, I find there is a much wider variety of tools needed to keep up with general property maintenance—especially when working with more perennials in the food forest and orchard. Here are some tools across an array of needs to jog our design-mind when decision-making for the diversified Permaculture homestead or small farm.

The Permaculture homesteader often finds they are working in the bush, clearing new land or the excess vegetation dropped by the food forest. In both cases, the **brush hook** (A) will come in handy to chop and drop material out of reach. A long-handled **grape vine billhook** (B) helps prune vines or even cut and drop out-of-reach grape clusters on an arbor. A **short-handle billhook** (C) is great for *chop-and-drop* in the food forest, where vegetation is cut back and left as *in situ* mulch on the ground. **Pruning shears** (D) are always needed for small limbs, while larger-diameter pruning can be done with these **Castellari Anvil loppers** (E). A **tree climbers saw** (F) is also handy for limbing. Maybe you just need **small snips** (G) for herb cutting or tomato suckering. Homesteader with a lot of woodland will employ a variety of **axes** (H) and **hatchets** (I) for felling trees and chopping firewood. Keep count of tree quantities in your stand with a **tree counter** (J) that logs numbers with a simple click. Long-handle and short-handle **hookeroons** (K) are used to snag and move logs and firewood when working in the bush. **Wedges** (L) are useful when felling trees to *wedge* into your cut (notch) and orient the felling direction and assist the splitting of the tree trunk, especially when felling a tree in a direction opposite to its lean. Or you may need a selection of good **peening hammers** (M) to keep your scythe blades ship-shape for cutting brush under your orchard trees. This **dry-wall square** (N) is great for fast measurement and adjustment of Permabed and hügelkultur mounds to make

16. Cleanup and Maintenance   203

them equidistant across their top and between. A **hand auger** (O) can assist when placing tree support stakes and small-livestock fencing (larger versions may be electric or gas-powered). These **metal stakes** (P) are great to lay out new plots by marking corners and other landscape features when doing earthworks on a property. A variety of different spades and shovels will always be needed for different tasks, from planting trees to applying compost. This **spade from Holden Tool Co.** (Q) has a curved blade that makes a perfectly round hole—great for lifting nursery perennials to pot them up for sale. This **hay rake** (R) is 40" wide and heavy duty enough for use in wide-bed-top growing that maximizes growing space

versus path space (see 123 Planting Method). Digging and other garden forks (like the **golden rose fork** [S]) are some of the most versatile tools on the Permaculture homestead—used for everything from carrot harvesting to subsoiling between perennials in an edible hedge. A **short-handle grub hoe** (T) does good work turning soil when terracing sloping land. Maybe you have bees to help pollinate the food forest and provide honey, wax, and more! In that case, many tools are needed, possibly a **smoker** (U) to help access the hives. Perhaps you are collecting nuts from the orchard floor with this **nut gatherer roller** (V). With so much variety of annual and perennial crops on the Permaculture homestead, you will want to have quality **wood plant label stakes** (W) and fresh permanent **markers** (X). Always keep all those fruit trees well supported with **support ties** (Y) when the branches strain and droop when loaded with fruit. You will probably need cold storage. A **Coolbot** (Z) coupled with an over-the-counter air conditioner is a good and popular solution (see feature in Cold Storage chapter). Most of these tools can be found through Holden Tool Company.

FOCUS
Fall Cover Cropping

Sometimes, all you need to get some cover cropping done is the seed, a scoop, and a good bag. This is the most traditional form of seeding. To honor that, I use this bag from my dad that is rooted in the tradition of the land.

The Work Corner

This "work corner" is an in-between space between the workshop and wash area, the farm office and the outdoor fields. It's the perfect place to house essentials that are routinely needed. A work corner should be easy to access so you can get what you need when you need it! This **corner** (A) exemplifies some items you may want to include for your own space and personnel management. It has some essentials for cleanup and other important safety items for work. A **clock** (B) can help people be prompt and keep on task. Spare **sun hats** (C) keep people covered on hot days, and spare warm **hats and gloves** (D) are stored nearby, too. Essential **supplies** (E) like extra toilet paper, towels, sanitizers, and hand soap refills are in view for replenishment. **Safety masks** (F), **protective gloves** (G), and a **first aid kit** (I) are in easy reach. Other safety items, like **hard hats** and **protective glasses** (H), are also available. A **fire extinguisher** (J) is mounted and labeled. A **cooler** (K) can hold cool water and food when headed into far fields or on other excursions, like seed collecting in the woodlot. **Cleanup essentials** (L) like brooms, dust pans, and snow shovels are kept by the main door so they are easily accessible. Large vacuums may be stored in a bigger area, but a small **shop vac** (M) is good to have handy to keep spaces tidy. Work corners serve many purposes, and safety, worker needs, and cleanup are an important organizational focus for work corners.

In the Tool Shop

Many growers forget to do basic maintenance of their tools. It is admittedly easy to use tools, put them away dirty (or not at all), and continue to use them until they wear out and require fixing. As **pro growers**, we should do more than just maintain; we can even engage in the modification of our favorite tools and to create entirely new tools better suited to our specific garden operations. Some activities that can occur in the shop include:

- **Making supplies**, (1) like these **oak row stakes** (A) for use as row markers. You can cut them to the perfect size to be strong and reusable, visible in the fields, and easy to write crop variety names and dates on (we also attach longer-lasting **metal labels** (B) to the larger row stakes for perennials). When painting the stakes, try using natural bristle **brushes** (C), like these from Sage Restoration, that capture paint into the middle of the brush for more precise work along the narrow surfaces of the stake. Painting the stakes with this Allback **linseed oil paint** (D) helps them last longer, and makes them even more visible and easier to write on with a paint marker.
- **Maintaining tools** (2) is best done with regular maintenance, including oiling wood handles with **boiled linseed oil** or **raw linseed oil** (E). *Note:* Many "boiled linseed oils" are *not* boiled and contain toxins; either way, always use rubber gloves. I prefer heavy-duty, **reusable rubber gloves** (F) because light-duty surgical gloves rip, and normal work gloves will get messy. Linseed oil is best applied with a **lint-free rag** (G), which can be made by cutting up old clothing. *Note:* Linseed oil is combustible. Used rags shouldn't be tossed aside into a random box or left out. Best to hang them to dry outside, somewhere out of direct sunlight.
- Tools can be cleaned after field use with wire brushes. I like this **brush** (H) with a scraper on one end for crusty material on shovel and hoe blades, or this stiff-bristle **scrub brush** (I) for cleaning handles and framing on tools. Rust is easily removed with **hand blocks** (J), using coarse, medium, and fine grit. Specialized **sharpeners** (K) for pruning shears and knives include whet stones, files, and diamond hones—all can be helpful for different applications.

16. Cleanup and Maintenance 207

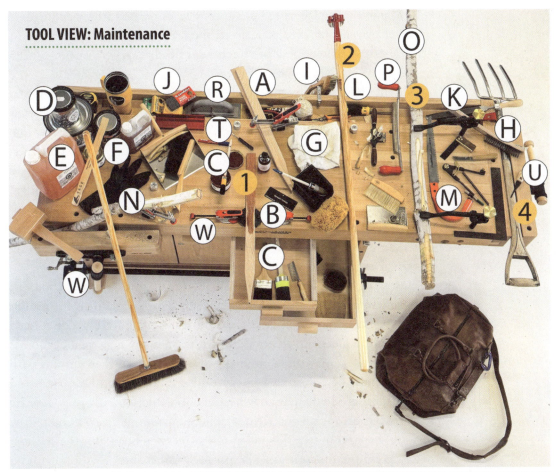

Tighten or replace any bolts or screws as needed with handheld nut drivers and cordless drivers; the latter is useful to clean and de-rust with a number of **attachment heads** (L).

Making new tools from scratch (3) is a great way to make use of extra and found materials, fine-tune tools to your needs, and build entirely new devices for your garden applications. We use a chainsaw or **handsaw** (M) for cutting our planted coppices of hardwoods for posts, small-dimension lumber, and **short-** (N) and **long-handled tools** (O). A **drawknife** (P) can help to shave down wooden staves for macro-shaping into tool handles, which can then be refined with a round **spokeshave** (Q). Then we can use a **block sander** (R) or an orbital sander to clean the surface before oiling, or painting the handle and attaching favorite tool heads or new custom designs.

Restoring and modifying old favorites (4) is a great way to spend time. This garden fork has harvested thousands of pounds of carrots for me but was abandoned one year in a garlic field that went fallow; it was discovered in an "archeological dig" three years later. Despite my high qualification as an archeologist from years of this happening (common when farming with students!), it is best to have a great cleaning and storing routine to avoid this. That being said, we always will have tools we need to restore. Maybe we buy classic tools at yard and farm sales for a fraction of the cost and restore them to working order. This fork had its tines scrubbed, but the stainless steel was pristine underneath; the handle was **brushed** (I) thoroughly to clean out as much dirt from the wood pores as possible, then sanded with coarse and fine **sandpaper** (R). The wood block in the Y-handle was removed by **grinding** (S) off rivets, and the inside was cleaned with a narrow **wire brush** (T) and sanded. (If you restore an old tool, think about carving in your initials with a **chisel** [U] to leave a record of its history.) The fork was then ready to oil with three or more coats of linseed oil, allowing each layer to soak in over 7–10 days. Next, the Y-handles' block will be glued in with a structural sealant like this penetrating **epoxy adhesive** (V), and a new hold drilled and bolted while being held in a **vice** (W). Maybe paint a nice red handle with **linseed oil paint** (D).

Other important maintenance tasks for maintenance include oiling moving parts with **lubricant** (X), soaking and de-rusting using **de-rusting**

agents or vinegar (Y) and baking soda, and keeping moisture at bay with an application of **mineral oil** (Z), linseed oil, and beeswax. *Note:* Tung, walnut, and other oils are used for maintaining tool handles, but these cost more than linseed oil.

FOCUS
Tool Maintenance in a Nutshell

When it comes to tool maintenance, it is all about daily cleaning, seasonal care, and occasional repair. It is also about knowing the best techniques to do the cleaning, care, and repair—and having the right supplies and … well yes, the best tools to do the job right! Here are some quick tips:

- Buy the right tools needed to help maintain your tools. You need *tools to care for tools*.
- Acquire and store the best supplies for maintaining your tools so you have what you need when you need it.
- You need a workshop space to do this work, preferably with a workbench, good lighting, places to store supplies and tools, and room for projects in progress.
- Cleaning dirt from tools should be done after each gardening day, not the next week or month.
- Cleaning rust is easily done with vinegar or a CLR soak if it builds up. But cleaning your tools and storing them dry prevents rust, and a foam square or steel wool with baking soda does an easy touch-up quickly.
- Oiling wooden handles should be done immediately if you buy tools with unfinished handles. You will need to do several rounds of oiling and letting them dry to let the oil soak in. This is especially true if using raw linseed oil (which is nontoxic compared to boiled linseed oil).
- Oiling metal parts, blades, rivets, and moving parts seasonally is important to keep them in good working order.
- Sharpening blades, big and small, is done at the end or start of the season before the gardening gets busy again. A sharp hoe blade slices fast through soil and weeds!
- Replacing worn parts and making minor repairs is best done in winter in a heated workshop.

210   *The Garden Tool Handboook*

TOOL VIEW: Using the Shop Work Bench

There are always **jobs** to do on the farm (A). **Vera** (B) is set up to get her jobs done. Everything on the **tabletop** (C) can be stored in the drawers of this Beaver Workbench from Canadian Woodworker. Essentials include a bench-sweeping set to keep the work surface tidy even when deep into projects, **hand saws**, **a compass**, **a square** (E), a **block sander** (G), bench stops, and vices. Your tea and small tools tuck nicely into place in the **tool well** (F), keeping you organized and alert while working. Supplies like **linseed oils** (H) can be kept in a cabinet below and brought out for routine maintenance and seasonal jobs like painting new **row marker stakes** (I). Ergonomic work occurs when all **supplies** (J) are nearby and within reach and your **project items** (K) are secured with vices (L) or placed ergonomically for efficient workflow. Make use of your **top drawer** (M) for most-needed items, like these natural bristle **brushes** (O) for efficient painting.

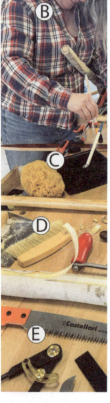

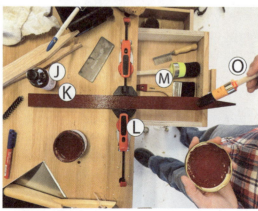

Peening Your Scythe

When your **scythe blade** (A) gets dull from use, it is time to peen its **edge** (B). This could be done by filing or grinding the edge—but that process sharpens by removing bits of metal. The removal of all those little bits can add up, reducing your scythe's weight and balance. **Peening** is the process of *re-edging* your scythe blade by hammering the metal back into shape and creating a work-hardened edge without removing any metal from the blade itself. This **scythe hammer** (C) has one convex face that concentrates the hammer blow onto a small area; it's used to compress the scythe blade metal when working thicker edges that need to be drawn out and thinned into their true and sharper shape. The hammer's more flattened face is used for wider overlapping strikes to improve the uniformity of the re-shaped scythe edge. This hammer is meant to be used with the peening anvil. This Austrian-made **SFX anvil** (D) has guides to help keep the scythe blade steady when working, which is helpful while developing the new skill of peening. *Note:* This peening anvil has an edge (not a flat face), so it works with the scythe hammer. You can peen with a normal **flat anvil**, but then you would need a specialized **hammer cross-pein** or **blacksmith hammer** (not pictured). This anvil is a **bar anvil** and would be used with the flat-faced scythe hammer. A medium grit (E) and fine grit (F) **synthetic stone** can be used to keep your blade sharp between peening sessions. The **Sandflex rust eraser** (G) is a great tool for keeping the blade clean and preparing it for peening.

Assembling Tools

Tool assembly needs are straightforward: a clean workspace, all the tool components, basic instructions, and the essential tools for putting the thing together. Some tools come with better instructions than others. When instructions are unclear, call the company and/or refer to photos and videos online for more details.

This tilther from Johnny's Selected Seeds comes with the **hood assembly** (A) already attached to the **carrying handle** (B), and these are attached to the **handlebars** (C) with **carriage bolts** (D). The square side of the carriage bolts goes to the outside of the handlebar and needs to be hammered in gently (E) so the square shoulder of the bolt is embedded in the wood. The **finger trigger** (F) is connected to the right handlebar and attached to the **trigger cord** (H), which runs down the handlebar through the **eye hooks** (G) that can be hand screwed into the provided holes. Your **cordless drill** (I) rests into the place allotted and is held firm with the **strap** (J). Then the **cable** (H) is run in order of a series of three marked points (1, 2, 3) to ensure it wraps around the trigger properly before being tightened by the **clamp** (L). Lift the tilther at the **crossbar** (M), and you are ready to set into the soil where the **gearbox** (N) is ready to power the **tine assembly** (O) using the power from a cordless drill.

16. Cleanup and Maintenance

TOOL VIEW: Zach's Storage and Organization System

Organization is essential on the homestead and farm. Here, I present some of my own tried-and-true methods and supplies. This bulk **bin organizer** (A) I picked up was discounted because it had damage to its corner—but it had endless possibilities. This wall-mount **panel rack** (B) can hold various **supply bins** (C), I find this very useful for irrigation pieces and nuts, bolts, and screws. Wall-mount **tool-hanging systems** (D) are abundant and useful, but not all of them can handle the heavier professional garden tools you might have. Clips, hooks, and large **carabiners** (E) can be handy to ceiling-mount larger items to get them out of the way. Screw-in tool **holders** (F) and hooks are very commonly used to keep tools handy. This **packout system** (G) by Milwaukee is an upgrade from the homemade version I used for years (a typical drawer system strapped to an industrial dolly). What I love about this system is its modular nature, but the **slender profile organizers** are what initially caught my eye. These are excellent for storing the thousands of ID tags that I have for the Edible Biodiversity Conservation Area's plant trials and any number of cases can be grabbed and carried into the field. On my farm, I also use practical **magnetic bars** (H) for easy tool wall organization, reused **milk crates** (I) for modular tool storage, the **classic toolbox** (J), which can be refurbished (K) and for which you can make your own **wooden tool trays** (L). **Tool buckets** (M) are good for job site projects (like building your own farm infrastructure), and **50-gal drums** (N) are very handy for stake storage. **Tool bags** (O) are my preferred field kits for transplanting and carrying other essentials, though **tool belts** (P) are also useful for carrying task-specific essentials. When **drill bits** (Q) come in a kit, keep these together.

FOCUS
Tool Walls

Keep your tools conveniently located. Greenhouses have their own needs and can have a specific **tool wall** (1). This one is built with a custom plywood tool **French cleat** holder system (2). Sometimes hanging items with **nails** (3) keeps items visible (and decorative, in this case) until needed. Supply **bins** (4) are always useful, and a custom-drilled screwdriver **holder** (5) is easy to make and helpful to keep screwdrivers organized. Sarah and Phillip at Gemüsezeit Altluneberg GbR farm in Germany have a place for every tool in their **tool corner** (6) to help them manage their intensive, 3,230 ft² (300 m²) garden and 5,600 ft² (520 m²) greenhouse for direct marketing of vegetable baskets and plant starts. Clip-style **toolbars** (7) are modular and allow you to adjust your spacing of tools as your systems evolve. **Alternating types of tools** (8) like hoes and shovels can help save space. However you slice it, the tool wall is about keeping what you need—like these Terrateck hoes—close at hand.

Above left: *Outdoor tool walls under a rain shelter and near the garden can be very convenient.*

Left: *Photo credit image 6: Gemüsezeit Altluneberg GbR.*

16. Cleanup and Maintenance   215

FOCUS
Tool Sheds

Another popular way to store tools is in a shed. They are most convenient to use when they are adjacent to the garden itself, which is especially helpful for those growing in community gardens or in larger plots where the garden may be hundreds of feet from the main buildings.

Roi unlocks the garden shed at the urban community garden where he can access tools supplied for gardeners to use.

FOCUS

Making the Most of Vertical Space for Workspace

There are many small tools and supplies that need to be stored. Our **seed library** (1) and **workshop** (2) both demonstrate good use of vertical space for work and storage.

Seed is an important supply that requires low humidity and a cool space. I have stored seed in small tote bins, pull-out drawer systems, etc. I have a lot of seed to store these days and cycling saved seed in and out when seed saving makes a good storage system important. I have evolved to using a **toolbox system** (A) with boxes of different sizes that allow storage of small seed packets in **pull-out drawers** (B) and larger bulk bags in **top-opening compartments** (C). Smaller toolboxes serve for **niche crops** (D). Our **seed library** (E) is a dedicated space in the barn that has some climate control. There is **shelving** (F) for storage above and a **workbench** (G) space for sorting seeds, calibrating **seeders** (H), and loading up the **seeding kit** (I) to head to the fields. Large bags of cover crop seed are stored in **totes** (J) on the slab floor.

In the next bay of the barn is our workshop space dedicated to DIY projects and maintenance. **Workbenches** (K) have additional **task lighting** (L), and access to power for important services like **air pressure** (M) to pump up **wheels** (N) or cart tires. Magnetic **toolbars** (O) help keep tools handy, and **peg boards** (P) are a good modular wall tool system. A **secondary workbench** (Q) allows the socket wrench set to lay open for use and has space for storing clamps, vices, and other items. **Bulk bins** (R) below store larger supplies, while **mid-level shelves** (S) store tool kits, and **high shelves** (T) store light supplies that are needed only seasonally **and** can be accessed with a ladder. A homemade heavy-framed **worktable** (U) can take a pounding when needed. A modular bin wall is good for **nuts**, **bolts**, **screws**, and such little items (V).

16. Cleanup and Maintenance

Finding Your Essentials

Most of us will find our essential tools naturally through circumstance and experience. We will acquire them as budgets allow and garden needs insist.

Essential Tools in Turkiye

To offer perspective on essential tools, I went to a Turkish bazaar in a small village to buy some tools. I was traveling while doing research for a book and working on building an edible ecological education site in the community. So I needed some tools to help get the work done. But I was curious about what was readily available and posed myself this question: "If I were to start homesteading on a small plot here in central Anatolia, what would I buy at market on this specific Tuesday?" Not everything I wanted was available, and something I didn't know I needed was. I think this experience provides some insight into what essential tool collections look like for different people in their own environments and situations.

   I found a great **spade** (A) which proved exceptionally strong (the shopkeeper smacked it against the ground several times to show the durability). But it was the wooden foot pad that he mounted further up the handle and the 1" diameter of its handle that impressed me for its common sense—previously I had never bought a spade with such an ergonomic and practically placed foot pad. I didn't find a broad hoe (but he said next week he would bring one), so my spade served to both break new ground and double-dig the garden plot. I also found a **rake** (B), not as wide as I typically like for bed preparation, but strong and practical for roughing in beds. I found a good **shovel** (C) to clear the path material and apply compost to bed tops. A **hay fork** (D) had a practical tine width for straw mulch and for manure moving. I also picked up a **hand hoe** (E) for between-row weeding. I would have liked a longer handle, though. This one would be more sensible for container gardening. A good **hatchet** (F) was immediately put to work by the shopkeeper to cut handles to the best length for me and fit them into the shovel and spade heads. This **hammer/crowbar** (G) is very practical on the homestead, as are the ever-needed **pipe wrench** (H) and **pipe cutter** (I), both of which I use for irrigation manifold and distribution piping installation. Medium and smaller **hammers** (J) are always needed for work on the farm. So are **whetstones and files** (K) for sharpening. These little artisan-made **hand saws** (L, M) caught my eye, and I picked up two. The smaller one proved to have a very fine saw blade for smaller cuts on small-diameter fruit branches, and the longer **saw** (M) served for larger-diameter work. A **pruning shear** (N) was also needed, along with a **pocket knife** (O) to cut greens and wild herbs (this was in April, so I saw many people were gathering greens and herbs). This **device** (P) launches a small explosive into ground vermin holes. (I saw, by the many holes in the fields, that this was needed. Populations were huge.) Otherwise, this heavy-duty metal **mouse trap** (Q) would keep populations low in the barn. A **lead rope** (R) reminded me of the need for animal husbandry to manage the larger fields and obtain nutrient sources. This **tent stake** (S) has many uses, including as a hose run for the garden path so the hose doesn't drag over the bed tops when someone is hand-watering smaller plots. A good **broom** (T) is always useful, as is a dustpan. **Work gloves** (V) were also purchased.

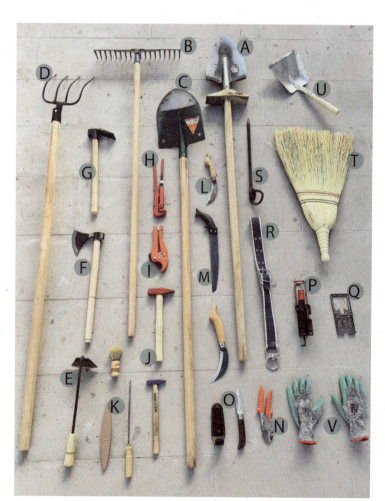

Again, I would have hoped for a broad hoe, and some long-handled weeding tools, as well as, of course, as some of the specialized tools I enjoy. But practically speaking, the items I found that day would be valuable for a start-up homesteader in Anatolia.

Essential Tools in Turkiye

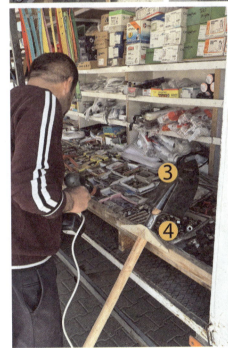

The shopkeeper had a supply of many different handles as **sanded staves** (1). These handles had the top rounded for hand comfort and the bottom shaved to fit tool heads right in the shop. He also had the heads for **many different tools** (2)—from spades and hoes to axes and hammers—to be mounded to the handles. Right there, in this roadside shop, the **head for my spade** (3) was mounted to the handle, and a special **foot pad** (4) was mounted—a bit higher up than you usually see—which I found quite nice to use. Using an **axe** (5), the **tool handle** (6) was trimmed to fit the **shovel** (7).

A spade, shovel, and rake were pretty much all I needed to build this brand-new garden plot. Garden hose sections fitted over three rake tines served as row markers.

Conclusion

The garden tool revolution is here to stay. When we take the right tools into our hands, we can solve our garden's weak links to make our home gardens, homesteads, and small farms much more productive, beautiful, and efficient. An understanding of the whole garden operation cycle and pairing each production stage with the right tools and techniques for your scale makes for a successful growing season year after year!

Well-made and carefully selected tools support the resilience and food security of our communities and the profitability of our farms. The resilience fostered by tool systems stems from our most ancient ancestors who picked up the first sticks, selected the first seeds, cultivated the first soils, and made their gardens grow! Tools in hand and intention in mind to grow great food is not only a profitable endeavor for any home gardener, homesteader, or small farmer but also provides a deep-rooted connection to what it means to be human.

Grow On!

—Zach

Resources

List of Tool Companies and Suppliers

Tools discussed in this book are manufactured and/or found through the following tool companies and suppliers.

BCS: Manufacturer and supplier of two-wheel tractors and attachments. bcsamerica.com

Black Swallow Living Soils: Supplier of soil mixes and fertility amendments. blackswallowsoil.com

Boot Strap Farmer: Manufacturer and supplier of potting systems and tools for greenhouses. bootstrapfarmer.com

Canadian Woodworker: Seller of quality woodworking benches and woodworking tools. canadianwoodworker.com

Dubois Agrinovation: Supplier of irrigation equipment, farm supplies, and tools. duboisag.com

Exaco: Distributor for cold frames and other infrastructure including greenhouse kits and home compost systems. exaco.com

Farmers Friend: Supplier for specialized market garden tools. farmersfriend.com

Growers and Co.: Tool company specializing in popular market garden tools and Canadian-made tools for market growers. growers.co

Growing Resources: Manufacturer and supplier of flame weeders of different sizes and accessories. flameweeders.com

High Mowing Seeds: Supplier of organic seeds and supplies. highmowingseeds.com

Holden Tool Company: Distributor of quality European hoes and other tools for gardening, forestry and small-scale farming. farmandgardentools.com

Jacto: Manufacturer of an array of backpack and equipment sprayer types and sizes useful for commercial and home growers. jacto.com

Johnny's Selected Seeds: Seed supplier and tool company with popular tool systems. johnnyseeds.com

Juwel: Manufacturer of cold frames in Austria. juwel.com/en/

Lapp Wagons: Manufacturer and supplier of wagons and carts. lappwagons.com

Meadow Creature: Maker of specialized garden tools and popular broadfork models. meadowcreature.com

Neptune's Harvest: Producer of quality fish-based and other fertilizers. neptunesharvest.com

Paperpot Company: Supplier of Paperpot system and market garden tools. paperpot.co

Preston Hardware: An independent hardware store with many tools needed for DIY jobs on the farm and tool maintenance. prestonhardware.com

Reforged Ironworks: Small-scale manufacturer of garden tools made from reused and re-forged iron. reforgedironworks.com

Sage Restoration: Supplier of woodworking and tool maintenance supplies and tools, like linseed oil and paints. sagerestoration.com

Scythe Works: Manufacturer and distributor of scythes, sickles, and other European tools. scytheworks.ca

SHW Tools: Manufactured of forged tools for garden and forestry in Germany. shw-fr.de

Sneeboer and Zn: Manufacturer of quality forged tools with a great array of types and sizes. sneeboer.com/en-us

Sproutbox Garden: Manufacturer and supplier of modular raised garden container systems. sproutboxgarden.com

Terrateck: Manufacturer and supplier of wheel hoe systems and tools for market growers from France. terrateck.com/en/

Two Bad Cats: Manufacturer and supplier of rolling dibblers and other essential garden tools. twobadcatsllc.com

Vital Soil Services: Tool supplier in Australia with specialized tools and services. vitalsoils.com.au

Index

Page numbers in *italics* indicate photographs and illustrations.

A

Alternate Maturity Patterning, 159, *159*
amendments, 151, *151*
ancestry, connection to, 78
animals, as garden tools, 7
attachment tool systems, 19, 21, *21*
avatars, 64–65, *65*

B

back-to-the-landers, 64
bagel trays, 129
Beaverland Farms, 146–147, *147*
bed-forming tool guilds, 51, *51*
best management practices, 44
between-row weeding, 166–168, *167*
bins, 19, 178, *178, 179*
biological pest management, 156
blind weeding, 164
breaking ground, 90, 93
Brisebois, Dan, 79
broadforks, 34, *34*

C

calendars, for planting schedule, 78–79
carrots, 112
carts, *37*, 174, *175*
caterpillar tunnels, 153–154
cell trays, *89*
Cleanup and Maintenance stage
 field cleanup, 198–199, *199*
 scythe peening, 211, *211*
 shop work benches, 210, *210*
 tool assembly, 212, *212*
 tool maintenance, 209
 tool organization systems, 213–216, *213–216*
 tool shops, 206–208, *207*
 work corners, 205, *205*
cold frames, for season extension, 180, *181*
Cold Storage stage
 CoolBot system, 192–193, *193*
 root cellars, 191–192, *191*
collars, 11–12, *13*
community gardens, 66, *66*, 109
complete tool systems, 52–53
compost, 151, *151*, 198–200, *199, 200*
Compost-a-path system, 200, *200*
Connecta system, 21, *21*
container gardening
 containers, 102–103, *103*, 105
 tool length, 104
 tools, 106, *107*
CoolBot system, 192–193, *193*
coppicing, for tool handles, 8, *8*
cover crops, 158–162, 204
crates, 178, *178, 179*
crop maintenance.
 See Garden Crop Maintenance stage
crop planning.
 See Garden Design and Crop Planning stage
Crop Weeding stage
 between-row, 166–168, *167*

Index

carts and wagons, 174, *175*
flame weeders, 114
in-row, 168–169, *169*
mulch and weed barrier, 172, *173*, 174
row spacing and hoe size, 164–165
scale and, 57
tool systems, 53, *53*
using wheel hoes, 169–171, *169*, 171
Cully Neighborhood Farm, 36–37

D

debris management, 198–200, *199*, *200*
decision-making, scale-based, 56–57, *58*
design
  forks, 31–34, *33*
  form for function, 22
  hoes, 23, *23*
  knives, 27, *27*
  rakes, 24, *24*
  shovels and spades, 25, *25*
  task-specific tools, 26, *26*
  tool systems, 51–55
  trowels, 28–30, *29*
  dibbling, 116, 118–119, *119*
direct-seeding, 81
DIY tools
  advantages and disadvantages of, 35–36
  category, 19
  drip roller, 36
  potting setup, *86*
  projects, 38, *38*, 41, *41*
  row marker, 39, *39–40*
  scaling-up setup, 88, *88*
  for season extension, 182–183, *182*
  seeding tables, *86*
drip rollers, 36
drip tape vs sprinklers, 141
dual-purpose tools, 19

E

ecosystem services, 19
Edible Biodiversity Conservation Area (EBCA), 6–7
environmental site assessment, 68–73
EPI (Education, Propagation, Inspiration) sites, 21
extensive garden management, 56–57

F

farmer's market tools, 194–195, *195–197*
fasteners, 12, *13*
fertilization
  application tools, 148–150, *149*
  compost, 151, *151*, 198–200, *199*, *200*
  management stages, 152
  supply organization, 155, *155*
field cleanup, 198–199, *199*
field labeling, 133, *133*
field recordkeeping, 162–163, *163*
field transplanting, 125–127, *127*, 129
Fine Seedbed Preparation stage
  finishing tools, 110, *110*
  pre-weeding tools, 112–113, *113*
  requirements based on crops, 108
  roughing in, 109
  succession bed preparation, 114–115
  weed management, 111–112, 114, 165

Fisheye Farms, 186–187, *187*
flame weeders, 114
forks, variations by task, 31–34, *33*
form for function
  about, 22
  forks, 31–34, *33*
  hoes, 23, *23*
  knives, 27, *27*
  rakes, 24, *24*
  shovels and spades, 25, *25*
  task-specific tools, 26, *26*
  trowels, 28–30, *29*
furrowing, 124–125, *124*

G

garden cleanup, 198–199, *199*
Garden Crop Maintenance stage
  cover crops, 158–159
  fertility tools, 148–150, *149*, 151, *151*
  fertilization management stages, 152
  field recordkeeping, 162–163, *163*
  mowing, 160–162, *161*
  pest management, 148–150, *149*, 152–153, 156, *157*
  suckering, 150
  supply organization, 155, *155*
  tool systems, 53, *53*
  trellising, 144–147, *145, 146*
  tunnels, 153–155
Garden Design and Crop Planning stage
  crop plan schedule, 77–78
  crop selection for environment, 156
  seasonal feedback in crop planning, 76
  seed orders records, 76–77
  using site analysis for design, 74–75
  using spreadsheets, 79–80
Garden Harvest stage
  crates, 178, *178, 179*
  harvest tools, 176–177, *177*
  tool systems, 53, *53*
  See also Post-harvest Handling and Curing stage
garden operation cycle
  about, 45–46
  production stages, 2, 47–50, *49*
  scaling-up, 62–63
  seasonal stages, 46, *46*
  successful growing and, 44
  weak links, 52
  See also individual stages
Garden Starts stage
  germination, 81
  growing out, 83
  hardening off, 83
  potting up, 82, 86–87, *86, 87*
  scaling-up, 88, *88*
  task flow, 83–84, *85*
garden tools. *See* tools
gardens, organisms and environment in, 6–7
Gemüsezeit Altluneberg GbR, 42, *43*
germinating, 81, 83–84, *85*
green manure, 158–162, *161*
greenhouses, 154
grips, 11, 14–15, *15*
ground covers, 70
growing out, 83–84, *85*
guild enterprises, 64
guilds, 51, *51*

H

handles, 8, *8*, 11, 14–15, *15*
hardening off, 83
harvest.

See Garden Harvest stage; Post-harvest Handling and Curing stage
Hedgeview Farm Organics, 86–87
high tunnels, 154
hilling, 124–125, *124*
hoes
    row spacing and, 164–165
    for soil type, 94–96, *95, 96*
    variations by task, 22–23, *23*
    for weeding, 165, *165*
home gardeners, 64
homesteaders, 64
Hori Hori, 9
humans
    avatars, 64–65, *65*
    as tools, 4, *6*

I
inputs, 19
in-row weeding, *167*, 168–169
Integrated Pest Management (IPM), 152–153
intensive garden management, 56–57
Irrigation stage
    design for specific needs, 142–143, *142, 143*
    distribution to irrigation, 138–139, *139*
    kits, 54–55, *55*
    manifold to garden row, 141
    pumping and distribution, 135–137, *137*
    scaling-up options, 134
    sprinklers vs drip tape, 141
    system parts, 135

J
Jagger, Chris, 188, *189*
Jang seeders
    advantages and disadvantages of, 133
    calibration, 122–123, *122, 123*
joints, 11, 14–15, *15*
Juniper Farm, 86

K
kits, 54
knives, 27, *27*

L
labels, 133, *133*
leeks, 42, *43*, 119, *120*
living organisms, 6–7, 19
long-handle tools, 18
low tunnels, 153–154

M
maintenance
    of tools, 209
    See also Cleanup and Maintenance stage; Garden Crop Maintenance stage
maps, 68–69, *69, 71*
market gardens, 1
market growers, 64
Marketing stage, farmer's market tools, 194–195, *195–197*
master crop data sheet, 117
measuring wheels, 91
medium-handle tools, 18
mowing, 160–162, *161*
mulch, 172, *173*, 174
multifunctional tools, 18–19
multi-row tools, 19

N
natural environment assessment, 68–73, *69, 71*
natural pest management, 156, *157*

Northern Liberties community
   garden, 66, *66*

O

123 Planting Method
   principle, 121, *121*
   rolling row marker, 39, *39–40*
   weeding, 164
Orto Vulcanico, 190, *190*

P

Paperpot system, 128, *128*
paths, composting in, 200, *200*
Permabeds. *See* Plot and Permabed
   Forming stage
Permaculture farms, 201–204, *201,
   203*
Permaplot patterns, 98, *98*
pest management
   application tools, 148–150, *149*
   Integrated Pest Management
      (IPM), 152–153
   natural, 156, *157*
   supply organization, 155, *155*
pesticides, 152
plant diversity, 6–7, *7*
planting.
   See Seeding and Planting stage
planting schedule, 77–78
plants, as garden tools, *7*
Plot and Permabed Forming stage
   architecture, 99, *99*
   in community garden, 66, *66*
   container gardening, 102–106,
      103, 107
   laying out design, 97
   Permaplot patterns, 98, *98*
   procedure, 99–100, *100, 101*
   tool systems, 51, *51*, 53, *53*

Post-harvest Handling and Curing
   stage
   cleaning and packing crops,
      184–185, *185*
   large volume system, 188, *189*
   tomato curing, 190, *190*
   workflow and DIY system,
      186–187, *187*
potting up
   about, 82
   procedure, 86–87, *87*
   task flow, 83–84, *85*
power tools, 19
pre-weeding, 111–113, *113,* 165
primary earthworks, 93
Primary Land Preparation stage
   breaking ground, 90, 93
   hoe choice, 94–96, *95, 96*
   measuring, 91
   tool systems, 53, *53*
   using two-wheel tractor, 92, *92*
production stages. *See* garden
   operation cycle; *individual stages*
pro-up phase, 3, 60

R

rakes, 23–24, *24*
records
   calendars, for planting schedule,
      78–79
   digital vs paper, 80
   field recordkeeping, 162–163, *163*
   master crop data sheet, 117
   seasonal garden notes, 76
   for seed orders, 76–77
refrigeration systems, 192–193, *193*
rolling row marker, 39, *39*
root cellars, 191–192, *191*
row cover, 153

Index

row marking
  DIY marker, 39, *39–40*
  methods, 118–119
  tools, *117, 119*
  types of, 116
rows, spacing and hoe size, 164–165

S

scale
  about, 56
  principles and decision-making, 56–57, *58*
scale phases, 3, 59, 60
scaling-up
  avatars, 64–65, *65*
  phase, 60
  production stages, 62–63
  tool systems, 61–62
  weak links and, 3
schedules
  crop plan schedule, 77–78
  using calendar, 78–79
scythe peening, 211, *211*
Season Extension stage
  options, *183*
  tools, *181*
  tunnels, 180–183, *182*
seasonal operation cycle.
  See garden operation cycle
seed orders records, 76–77
seedbeds.
  See Fine Seedbed Preparation stage
seeders, 129–131, *130, 132*
Seeding and Planting stage
  field labeling, 133, *133*
  field trays, 129
  furrowing, 124–125, *124*
  row organization, 116–121
  seeders, *130, 132,* 133
  seeding, 129–131, *131*
  tool systems, 53, *53*
  transplanting, 125–127, *127*
seeds.
  See Garden Design and Crop Planning stage; Garden Starts stage; Seeding and Planting stage
shop work benches, 210, *210*
short-handle tools, 18
shovels, variations by task, 25, *25*
Site Analysis and Sampling stage
  assessment of natural environment, 68–73
  site assessment tools, *69, 71*
small-scale farmers, 64
soil
  as garden tools, 7
  hoe design and, 94–96, *95, 96*
  in pest management, 156
soil testing, 72–73
spades
  architecture of, 11, *11*
  variations by task, 25, *25*
specialized tool components, 16, *16–17*
sprayers, 148–150, *149*
spreadsheets, for crop plans, 79–80
sprinklers vs drip tape, 141
stale seedbeds, 111–112, 165
start-up phase, 3, 60
static scale, 60
stationary tools, 19
succession bed preparation, 114–115
suckering, 150
sun exposure assessment, 70, 72
supplies, 19
supports and trellising, 144–147, *145, 146*
synthetic weed barrier, 172, *173,* 174

systematic growing, 2

T
task-specific tools, form and function, 26, *26*
technique, in best management success, 45
tilthers, 212, *212*
timing, in best management success, 45
tine family. *See* rakes
tomatoes
   curing, 190, *190*
   in greenhouses, 42, *43*
tool guilds, 51, *51*
tool kits, 54
tool organization systems, 213–216, *213–216*
tool sheds, 215, *215*
tool shops, 206–208, *207*, 210, *210*
tool systems
   about, 20
   attachment systems, 19, 21, *21*
   development of, 52–53
   guilds, 51, *51*
   kits, 54–55
   scaling-up, 61–62
   for several production stages, *53*
tool walls, 214, *214*
tools
   architecture of, 11, *11*
   in best management success, 45
   essential, 217–218, *218, 219*
   history of, 4–5
   longevity of, 67
   organisms and environment as, 6–7
   selection of, 221
   traditional, 9, 10, 120
   type categories, 18–19

See also design
Tourne-sol Cooperative Farm, 79
traditional tools, 9, 10
transplanting, 125–127, *127*
   See also Seeding and Planting stage
trays, 19, *89*, 129
trellising, 144–147, *145, 146*
trowels, 28–30, *29*
tunnels
   DIY hoops, 182–183, *182*
   for season extension, 180–182, *181*
   types of, 153–155
Turkiye, essential tools, 217–218, *218, 219*
two-wheel tractors, 92–93, *92*

U
urban community gardens, 66, *66*, 109
urban growers, 64

W
wagons, 174, *175*
wash stations, 37
watering. *See* Irrigation stage
weak links, 45, 52, 62–63
weed barrier, 172, *173*, 174
weed management
in seedbed preparation, 111–113, *113*, 165
See also Crop Weeding stage
weed sweep, 164
wheel hoes, 169–171, *171*
wheel tools, 18
work corners, 205, *205*

Z
Zipperbed method, 174
Zurich community garden, 109

About the Author

Zach Loeks is an edible ecosystem designer, innovative researcher, and Permaculture grower. He helps homes, farms, schools, and municipalities create more edible and diverse landscapes. He is the author of *The Permaculture Market Garden, The Two-Wheel Tractor Handbook,* and *The Edible Ecosystem Solution.* Zach lives between two research projects: his award-winning farm in Cobden, ON, and his urban homestead in Winnipeg, Manitoba, and he translates his innovations into practical and empowering solutions through online offerings. Learn more at www.zachloeks.com.

Praise

This book is dedicated to the women farmers in my life: Deva, Kathleen, Carol, Gail, Vera, and my daughters Dayvah and Rainah Loeks. May your tools be sharp and your home gardens lush!

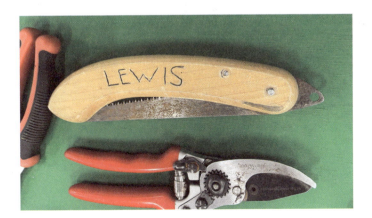

ABOUT NEW SOCIETY PUBLISHERS

 New Society Publishers is an activist, solutions-oriented publisher focused on publishing books to build a more just and sustainable future. Our books offer tips, tools, and insights from leading experts in a wide range of areas.

We're proud to hold to the highest environmental and social standards of any publisher in North America. When you buy New Society books, you are part of the solution!

- This book is printed on **100% post-consumer recycled paper,** processed chlorine-free, with low-VOC vegetable-based inks (since 2002).
- Our corporate structure is an innovative employee shareholder agreement, so we're one-third employee-owned (since 2015).
- We've created a Statement of Ethics (2021). The intent of this Statement is to act as a framework to guide our actions and facilitate feedback for continuous improvement of our work.
- We're carbon-neutral (since 2006).
- We're certified as a B Corporation (since 2016).
- We're Signatories to the UN's Sustainable Development Goals (SDG) Publishers Compact (2020–2030, the Decade of Action).

At New Society Publishers, we care deeply about *what* we publish—but also about *how* we do business.

To download our full catalog, sign up for our quarterly newsletter, and learn more about New Society Publishers, please visit newsociety.com.

ENVIRONMENTAL BENEFITS STATEMENT

New Society Publishers saved the following resources by printing the pages of this book on chlorine free paper made with 100% post-consumer waste.

TREES	WATER	ENERGY	SOLID WASTE	GREENHOUSE GASES
44	3,500	18	150	18,800
FULLY GROWN	GALLONS	MILLION BTUs	POUNDS	POUNDS

 Environmental impact estimates were made using the Environmental Paper Network Paper Calculator 4.0. For more information visit www.papercalculator.org